COLLINS
DIY
GUIDE

WEATHERPROOFING & INSULATION

J A C K S O N · D A Y

HarperCollins*Publishers*

Published by
HarperCollins Publishers
London

**This book was created exclusively
for HarperCollins Publishers by
Jackson Day Jennings Ltd
trading as Inklink.**

Conceived, edited and designed
by Jackson Day Jennings Ltd
trading as Inklink.

Text
Albert Jackson
David Day

Editorial director
Albert Jackson

Text editors
Diana Volwes
Peter Leek

Executive art director
Simon Jennings

Design and art direction
Alan Marshall

Additional design
Amanda Allchin

Production assistant
Simon Pickford

Illustrations editor
David Day

Illustrators
Robin Harris
David Day

Additional illustrations
Brian Craker
Michael Parr
Brian Sayers

Photographers
David Day
Peter Higgins
Simon Jennings
Neil Waving
Shona Wood

Picture researchers
David Day
Anne-Marie Ehrlich
Hugh Olliff

Proofreaders
Mary Morton
Alison Turnball

For HarperCollins
Robin Wood – Managing Director
Polly Powell – Editorial Director
Bridget Scanlon – Production Manager

First published in 1988
This edition published in 1995
Reprinted 1996

Most of the text and illustrations in
this book were previously published in
Collins Complete DIY Manual

ISBN 0 00 412764 1

Copyright © 1988, 1995
HarperCollins Publishers

The CIP catalogue record for this
book is available from the British
Library

**Text set in Univers Condensed
and Bodoni**
by Inklink, London

Imagesetting by
TD Studio, London

Colour origination by
Colourscan, Singapore

Printed and bound
in Hong Kong

Picture sources

Key to photographic credits
L = Left, R = Right, T = Top,
TL = Top left, TR = Top right,
C = Centre, UC = Upper centre,
LC = Lower centre, CL = Centre Left,
CR = Centre right, B = Bottom,
BL = Bottom left, BC = Bottom centre,
BR = Bottom right

Blue Circle Industries PLC:
29, 35B
Cement & Concrete Association:
39C, 39B
David Day: 20
Peter Higgins: 45
Simon Jennings:
10, 15L, 32, 34, 35T, 35C, 41
Rentokil: 15R
Neil Waving: 16, 21
Shona Wood: 39

CONTENTS

Cross-references
Since there are few DIY projects that do not require a combination of skills, you might have to refer to more than one section of this book. The list of cross-references in the margin will help you locate relevant sections or specific information related to the job in hand.

DRAUGHT-PROOFING

SEE ALSO
Details for:
Ventilation   69-76

Seal off the major draughts around windows and doors. For a modest outlay, draughtproofing provides a substantial return in terms of comfort and economy.

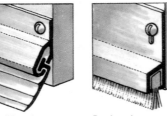

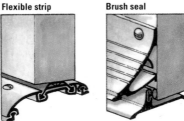

Flexible strip

Brush seal

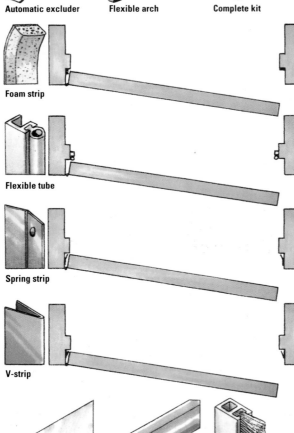

Automatic excluder   Flexible arch   Complete kit

Foam strip

Flexible tube

Spring strip

V-strip

Spring or V-strip

Compressible strip
Use for the central meeting rails.

Brush seal
The sashes slide past a brush seal.

Threshold draught excluders

If the gap between the door and the floor is very large, so that it admits fierce draughts, use a threshold excluder to seal the gap.

Flexible-strip excluder

The simplest form of threshold excluder is a flexible strip of plastic or rubber which sweeps against the floorcovering to form a seal. This kind of excluder is rarely suitable for exterior doors and quickly wears out. However, it is inexpensive and easy to fit. Most types work best on smooth flooring.

Brush seal

A nylon-bristle brush set into either a metal or plastic extrusion acts as a draught excluder. It is suitable for slightly uneven or textured floorcoverings, and can be fitted to hinged and sliding doors.

Automatic excluder

A plastic strip with its extruded clip is spring-loaded, so it lifts from the floor as the door is opened. As you close the door, the excluder is pressed against the floor by a stop screwed to the frame.

Flexible-arch excluder

An aluminium extrusion with a vinyl arched insert presses against the bottom edge of the door. If you plan to fit one on an external door, buy a version that has underseals to prevent rain from seeping beneath.

Complete door kit

The best solution for an outside door is a kit combining an aluminium weather trim to shed rainwater, which is fitted to the door, and a weather bar with a tubular rubber or plastic draught excluder for screwing to the threshold.

Weatherstripping door edges

Any well-fitting door requires a gap of 2mm (1/16in) at top and sides so that it can be operated smoothly. However, a gap this large wastes heat.

Foam strip

The most straightforward excluder is a self-adhesive foam-plastic strip, which you stick around the rebate: the strip is compressed by the door, forming a seal. Don't stretch foam excluders when applying them, as that reduces their efficiency. The door may be difficult to close at first, but the excluder soon adjusts to fit.

Flexible-tube excluder

A small vinyl tube held in a plastic or metal extrusion is compressed to fill the gap around the door. The cheapest versions have an integrally moulded flange, which can be stapled to the doorframe, but they are not as neat.

Spring-strip excluder

A thin metal or plastic strip that has a sprung leaf is pinned or glued to the doorframe. The top and closing edges of the door brush past the leaf, sealing the gap. The hinged edge compresses it.

V-strip excluder

A variation on the spring strip, the leaf is bent right back to form a V-shape. Most types are cheap and unobtrusive.

Draughtproofing sealant

In order to seal the gaps, a bead of flexible sealant is squeezed onto the door stop; a low-tack tape is applied to the surface of the door to act as a release agent. As the door is closed, it flattens the bead, filling the gaps perfectly. When the sealant has set, the parting layer of tape is peeled from the door. A flexible tubing, bonded with sealant, will compensate for movement.

Sealing a window

Hinged casement or pivot windows can be sealed with any of the draught excluders suggested for fitting around the edge of a door, but draughtproofing a sliding sash window presents a more complex problem.

The top and bottom closing rails of a sash window can be sealed with any form of compressible excluder.

The sliding edges of a sash admit fewer draughts, but they can be sealed with a brush seal fixed to the frame – inside for the lower sash, outside for the top one.

A springy V-strip or a compressible plastic strip can be used to seal the gap between the sloping faces of the horizontal meeting rails of a traditional sash window. Use a blade seal for square meeting rails.

DAMP
CAUSES

TYPES OF

DAMP

SEE ALSO
Details for:
Wet and dry rot   15

The symptoms of damp can be most distressing both in terms of your health and the condition of your home. Try to locate the source of the problem as quickly as possible, before it promotes its even more damaging side effects – wet and dry rot. Unfortunately, this is sometimes easier said than done, as one form of damp may be obscured by another, or may appear in an unfamiliar guise. The three main categories to eliminate are penetrating damp, rising damp and condensation.

Principal causes of penetrating damp
1 Broken gutter
2 Leaking downpipe
3 Missing roof tiles
4 Damaged flashing
5 Faulty pointing
6 Porous bricks
7 Cracked masonry
8 Cracked render
9 Blocked drip groove
10 Defective seals around frames
11 Missing weatherboard
12 Bridged cavity

Principal causes of rising damp
● Missing DPC or DPM
● Damaged DPC or DPM
● DPC too low
● Bridged DPC
● Earth piled above DPC

Penetrating damp

Penetrating damp is the result of water permeating the structure of the house from outside. The symptoms only occur during wet weather. After a few dry days, the damp patches dry out, often leaving stains.

As isolated patches are caused by a heavy deposit of water in one area, you should be able to pinpoint their source fairly accurately. General dampness usually indicates that the wall itself has become porous, but it could equally well be caused by some other problem.

Penetrating damp occurs most frequently in older homes, which have solid walls. Relatively modern houses built with a cavity between two thinner brick skins are less likely to suffer from penetrating damp, unless the cavity is bridged in one of several ways.

Rising damp

Rising damp is caused by water soaking up from the ground into the floors and walls of the house. Most houses are protected by an impervious barrier built into the walls and under concrete floors so that water cannot permeate above a certain level.

If the damp-proof course (DPC) in the walls or the membrane (DPM) in a floor breaks down, water is able to seep into the upper structure. Alternatively, there may be something forming a bridge across the barrier so that water is able to flow around it. Some older houses were built without a DPC.

This type of damp is confined to solid floors and the lower sections of walls. It is a constant problem – even during dry spells – and becomes worse with prolonged wet weather.

DPC in a solid wall
A layer of impervious material is built into a joint between brick courses, 150mm (6in) above the ground.

DPC and DPM in a cavity-wall structure
The damp-proof membrane in a concrete floor is linked to the DPC protecting the inner leaf of the wall. The outer leaf has its own damp-proof course.

7

DAMP: PRINCIPAL CAUSES

PENETRATING DAMP: PRINCIPAL CAUSES

CAUSE	SYMPTOMS	REMEDY
Broken or blocked gutter Rainwater overflows, typically at the joints of old cast-iron gutters, and saturates the wall directly below, preventing it from drying out normally.	Damp patches appearing near the ceiling in upstairs rooms, and mould forming immediately behind the leak.	Clear leaves and silt from the gutters. Repair the damaged gutters, or replace a faulty system with maintenance-free plastic guttering.
Broken or blocked downpipes A downpipe that has cracked or rusted douses the wall immediately behind the leak. If leaves lodge behind the pipe at the fixing brackets, that will eventually produce a similar effect.	An isolated patch of damp, often appearing halfway up the wall. Mould growth behind the downpipe.	Repair the cracked or corroded downpipe; or replace it, substituting a maintenance-free plastic version. Clear the blockage.
Loose or broken roof tiles Defective tiles allow rainwater to penetrate the roof.	Damp patches appearing on upstairs ceilings, usually during a heavy downpour.	Replace the faulty tiles, renewing any damaged roofing felt.
Damaged flashing The junction between the roof of a lean-to extension and the side wall of the house, or around a chimney stack emerging from the roof, is sealed with flashing strips (usually of lead or zinc) or sometimes with a mortar fillet. If the flashing or fillet cracks or parts from the masonry, water trickles down inside the building.	Damp patch on the ceiling extending from the wall or chimney breast; also on the chimney breast itself. Damp patch on the side wall near the junction with the lean-to extension; damp patch on the lean-to ceiling itself.	If the existing flashing appears to be intact, refit it securely. If it is damaged, replace it using similar material or a self-adhesive flashing strip.
Faulty pointing Ageing mortar between bricks in an exterior wall is likely to crack or fall out; water is then able to penetrate to the inside of the wall.	Isolated damp patches or sometimes widespread dampness, depending on the extent of the deterioration.	Repoint the joints between bricks, then treat the entire wall with water-repellent fluid.
Porous bricks Bricks in good condition are weatherproof, but old, soft bricks become porous and often lose their faces. As a result, the whole wall is eventually saturated, particularly on an elevation that faces prevailing winds, or where a fault with the guttering develops.	Widespread damp on the inner face of exterior walls. A noticeable increase in damp during a downpour. Mould growth appearing on internal plaster and decorations.	Repair bricks that have spalled, and waterproof the exterior with a clear water-repellent fluid.
Cracked brickwork Cracks in a brick wall allow rainwater (or water from a leak) to seep through to the inside face.	An isolated damp patch – on a chimney breast, for example, due to a cracked chimney stack.	Fill cracked mortar and replace damaged bricks.
Defective render Cracked or blown render encourages rainwater to seep between the render and the brickwork behind it. The water is prevented from evaporating and so becomes absorbed by the wall.	An isolated damp patch, which may become widespread. The trouble can persist for some time after rain ceases.	Fill and reinforce the crack. Hack off extensively damaged or blown render and patch it with new sand-cement render, then weatherproof the wall by applying exterior paint.
Damaged coping If the coping stones on top of a roof parapet are missing or the joints are open, water can penetrate the wall.	Damp patches on ceiling, near to the wall immediately below the parapet.	Bed new stones on fresh mortar and make good the joints.

PENETRATING DAMP: PRINCIPAL CAUSES

CAUSE	SYMPTOMS	REMEDY
Blocked drip groove Exterior windowsills should have a groove running longitudinally on the underside. When rain runs under the sill, the water falls off at the groove before reaching the wall. If the groove is bridged by layers of paint or moss, the water soaks the wall behind.	Damp patches along the underside of a window frame. Rotting wooden sill on the inside and outside. Mould growth appearing on the inside face of the wall below the window.	Rake out the drip groove. Nail a batten to the underside of a wooden sill to deflect drips.
Failed seals around windows and doorframes Timber frames often shrink, pulling the pointing from around the edges so that rainwater is able to penetrate the gap.	Rotting woodwork and patches of damp around the frames. Sometimes the gap is obvious where mortar has fallen out.	Repair the frame, and seal around the edges with mastic.
No weatherboard An angled weatherboard across the bottom of a door should shed water clear of the threshold and prevent water running under the door.	Damp floorboards just inside the door. Rotting at the base of the doorframe.	Fit a weatherboard, even if there are no obvious signs of damage. Repair rotted wood at the base of the doorframe.
Bridged wall cavity Mortar inadvertently dropped onto a wall tie connecting the inner and outer leaves of a cavity wall allows water to bridge the gap.	An isolated patch of damp appearing anywhere on the wall, particularly after a heavy downpour.	Open up the wall and remove the mortar bridge, then waterproof the wall externally with paint or clear repellent.

RISING DAMP: PRINCIPAL CAUSES

CAUSE	SYMPTOMS	REMEDY
No DPC or DPM If a house was built without either a damp-proof course or damp-proof membrane, the walls are able to soak up water from the ground.	Widespread damp up to about 1m (3ft) above skirting level. Damp concrete floor surface.	Fit a new DPC or DPM.
Damaged DPC or DPM If the DPC or DPM has deteriorated, water will penetrate at that point.	Damp at skirting level (possibly isolated but spreading).	Repair or replace the DPC or DPM.
DPC too low If the DPC is lower than the necessary 150mm (6in) above ground level, heavy rain is able to splash above the DPC and soak the wall surface.	Damp at skirting level, but only where the ground is too high.	Lower the level of the ground outside. If it's a path or patio, cut a 150mm (6in) wide trench and fill with gravel, which drains rapidly.
Bridged DPC If exterior render has been taken below the DPC or mortar has fallen within a cavity wall, moisture is able to cross over to the inside.	Widespread damp at and just above skirting level.	Hack off render to expose the DPC. Remove several bricks and rake out debris from the cavity.
Debris piled against wall A flower bed, rockery or area of paving built against a wall will bridge the DPC. Building material and garden refuse left there will also act as a bridge.	Damp at skirting level in area of bridge only, or spreading from that point.	Remove the earth, paving or debris and allow the wall to dry out naturally.

SEE ALSO

Details for:
Bridged cavity	10
Drip batten	10
Sealing frames	10
New DPC	12–13
New DPM	14, 30
Repairing frames	17, 25
Weatherboard	27
Waterproofing bricks	34

DPC too low

Render bridges DPC

Earth piled over DPC

TREATING DAMP

1 Water drips to ground

2 A bridged groove

3 Drip moulding

Apply mastic with an applicator gun

CURING DAMP

Remedies for different forms of damp are suggested in the charts on the previous pages; where damp conditions are attributable to factors such as poor ventilation or deteriorating decoration, you will find detailed remedies in other sections of the book. The information below supplements these instructions by providing advice on measures relating solely to the eradication of damp.

Waterproofing walls

Applying a water repellent to the outside of a wall not only prevents water infusion but also reduces the possibility of interstitial condensation. This occurs when water vapour from the inside of the house penetrates an external wall until it reaches the damp, colder interior of the brickwork, where the vapour condenses. The moisture eventually migrates back to the inner surface of the wall, causing stains and mould.

Proprietary damp-proofing liquids are available for painting the inside of walls, but they should be considered a temporary measure only, as they do not treat the source of the problem.

Remove wallcoverings and make sure that the wall surfaces are sound and clean. Treat any mould growth with a fungicide. Apply two full brushcoats over an area appreciably larger than the present extent of the damp. Once the wall is dry, you can decorate it with paint or a wallcovering.

If any of your walls show signs of efflorescence, apply the appropriate treatment, then paint with heavy-duty moisture-curing polyurethane.

Providing a drip moulding

Because water cannot flow uphill, a drip moulding on the underside of an external windowsill forces rainwater to drip to the ground before it reaches the wall behind (**1**). When redecorating, scrape out old paint or moss from drip grooves before it forms a bridge (**2**).

If a wooden windowsill does not have a pre-cut drip groove, it is worth adding a drip moulding by pinning and gluing a 6mm (¼in) square hardwood strip 35mm (1½in) from the front edge of the sill (**3**). Paint or varnish the drip moulding to match the sill itself.

Sealing around window frames

Scrape out old or loose mortar from around the frame. Fill deep gaps with expanding-foam filler or rolled paper, then seal all around the frame with a flexible mastic. Mastic is available in cartridges, some designed for use with an applicator gun. Cut the end off the nozzle of the cartridge and run it along the side of the frame to form an even continuous bead. If the gap's very wide, fill it with a second bead once the first has set. Most sealants form a skin and can be overpainted after a few hours, but are waterproof without painting.

Bridged cavity

A bridged wall cavity allows water to cross over to the inner leaf. The easiest way to deal with it is simply to apply a water repellent to the outer surface.

However, this does not address the cause, which may lead to further dampness in the future. When it is convenient, during repointing perhaps, remove two or three bricks from the outside in the vicinity of the damp patch by chopping out the mortar around them. Use a small mirror and a torch to inspect the cavity. If you locate mortar lying on a wall tie, rake or chip it off with an opened wire coat hanger or a metal rod, then replace the bricks.

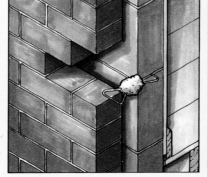
Exposing a bridged wall tie
Remove a few bricks in order to rake or chip the mortar from a wall tie.

Air carries moisture as water vapour, but its capacity depends on its temperature. As air becomes warmer, it absorbs more water, rather like a sponge. When water-laden air comes into contact with a surface that is colder than itself, it cools until it can no longer hold the water it has absorbed and (just like a sponge being squeezed) it condenses, depositing water in liquid form onto the surface.

Conditions for condensation

The air in a house is normally warm enough to hold water without reaching saturation point, but a great deal of moisture is also produced by cooking and using baths and showers, and even by breathing. In cold weather, when the low temperature outside cools the external walls and windows below the temperature of the heated air inside, all the extra water runs down window panes and soaks into the wallpaper and plaster. Matters are made worse in the winter by sealing off windows and doors so that fresh air cannot replace humid air before it condenses.

Damp in a fairly new house that is in good condition is almost invariably due to condensation.

The root cause of condensation is rarely simple, as it is a result of a combination of air temperature, humidity, poor ventilation and thermal insulation. Tackling just one of these problems in isolation may transfer condensation elsewhere or even exaggerate the symptoms. However, the chart opposite lists major factors that contribute to the total problem.

Condensation appears first on cold glazing

CONDENSATION: PRINCIPAL CAUSES

CAUSE	SYMPTOMS	REMEDY
Insufficient heat In cold weather the air in an unheated room may be close to saturation point. (Raising the temperature increases the ability of the air to absorb moisture without condensing.)	General condensation.	Heat the room – but do not use an oil heater (which produces moisture).
Oil heaters An oil heater produces as much water vapour as the paraffin it burns, causing condensation to form on cold windows, exterior walls and ceilings.	General condensation in rooms where oil heaters are used.	Substitute another form of heating.
Uninsulated walls and ceilings Moist air readily condenses on cold ceilings and exterior walls.	Widespread damp and mould. The line of ceiling joists is picked out because mould grows less well along the joists, which are relatively warm.	Install efficient loft insulation and/or line the ceiling with insulating tiles or polystyrene lining. Alternatively, apply anti-condensation paint.
Cold bridge Even when a wall has cavity insulation, there can be a cold bridge across the lintel over windows and the solid brick down the sides.	Damp patches or mould surrounding the window frames.	Line the walls and window reveals with expanded-polystyrene sheeting or foamed polyethylene.
Unlagged pipes Cold-water pipes attract condensation. The problem is often wrongly attributed to a leak when water collects and drips from the lowest point of a pipe run.	A line of damp on a ceiling or wall, following the pipework. An isolated patch on a ceiling, where water drops from plumbing. Beads of moisture on the underside of a pipe.	Insulate your cold-water pipes, either with plastic-foam lagging tubes or with mineral-fibre wrapping.
Cold windows When exterior temperatures are low, windows usually show condensation before any other feature, because the glass is thin and they are constantly exposed to the elements.	Misted window panes, or water collecting in pools at the bottom of the glass.	Double-glaze your windows. If condensation occurs inside a secondary system, place some silica-gel crystals (which absorb moisture) in the cavity between the panes.
Sealed fireplace If a fireplace opening is blocked up, the air trapped inside the flue cannot circulate and therefore condenses on the inside, eventually soaking through the brickwork.	Damp patches appearing anywhere on the chimney breast.	Ventilate the chimney by inserting a grille or airbrick at a low level in the part of the fireplace that has been blocked-up. Treat the chimney breast with damp-proofing liquid.
Loft insulation blocking airways If loft insulation blocks the spaces around the eaves, air cannot circulate in the roof space, and condensation is able to form.	Widespread mould affecting the timbers in the roof space.	Unblock the airways and, if possible, fit a ventilator grille in the soffit or install tile/slate vents.
Condensation after building or repairs If you have carried out work involving new bricks, mortar and especially plaster, condensation may be the result of these materials exuding moisture as they dry out.	General condensation affecting walls, ceiling, windows and solid floors.	Wait for the new work to dry out, then review the situation before decorating or other treatment.

SEE ALSO

● **Anti-condensation paint**
This paint contains minute hollow glass beads that act as insulators, as well as a fungicide to inhibit mould growth. It can be overpainted with emulsion to suit your colour scheme.

11

INSTALLING A DAMP-PROOF COURSE

When an old damp-proof course (DPC) has failed, or where none exists, the only certain remedy is to insert a new one. Of the options available, chemical injection is the only method you should attempt yourself. Even then, consider whether it is cost-effective in the long run. Rising damp can lead to other expensive repairs unless it is completely eradicated, so hiring a reputable company may prove to be a wise investment. (They normally provide a 30-year guarantee.) Ask for a detailed specification – known as an Agrément certificate – to ensure that the work is carried out to approved standards. Also, check that the guarantee is fully covered by insurance in case the company goes out of business.

Checking for rising damp

There's no substitute for a professional survey to determine the cause of rising damp, but you can use an inexpensive electronic moisture meter to check the condition of your walls.

Working on the inside, take readings at regular intervals along the entire length of a wall, not just in one spot. Systematically check an area extending from floor level to about 1m (3ft) above the floor. If rising damp is present, the meter should indicate a high moisture reading that drops sharply due to natural evaporation above that level. Penetrating damp or condensation tend to show up as isolated patches or even as dampness that extends right up the wall. Even if you suspect rising damp, check that there is nothing bridging a perfectly sound damp-proof course before committing yourself to major works on the DPC itself.

It is sometimes possible to detect symptoms of rising damp even after the installation of a new DPC. This is due to old salt-contaminated plaster, which should be removed and replaced with special renovating plaster.

Use a moisture meter to test for rising damp

A physical DPC

A traditional DPC consists of a layer of impervious material built into the wall at about 150mm (6in) – or two to three brick courses – above ground level. It is possible to install a DPC in an existing wall by cutting out a mortar joint with a chain saw or a grinding disc. Copper sheet, polyethylene or bituminous felt is then inserted and the joint wedged and filled with fresh mortar. Experience is needed in order to avoid weakening the wall, and there's always a risk of cutting into a pipe or electric cable. Although a physical DPC is expensive to install, it is considered the most reliable method.

Electro-osmosis

This method utilizes the principle that a minute electrical charge will prevent water rising by capillary action. A length of titanium wire is inserted in a continuous chase cut all round the building; anode points bent in the wire are then inserted into holes drilled in the masonry at regular intervals. The wire is connected to an earthing rod buried in the ground, and the system's power unit plugs into a standard 13amp socket. The holes and chase are filled with mortar to protect the wire. This type of system can be placed internally or externally, and has to be installed by a professional fitter.

Porous tubes

Porous clay tubes are inserted into a row of closely spaced holes to increase the rate of evaporation so that moisture will not rise to too high a level. This is a simple and cheap method.

A physical DPC A joint is removed to insert an impervious layer.

Electro-osmosis A copper electrode is planted in the wall.

INJECTING A CHEMICAL DPC

The most widely practised method is to inject a waterproofing chemical, usually silicone-based, to form a continuous barrier throughout the thickness of the wall. Suitable for brick or stone walls up to 600mm (2ft) thick, it is straightforward to install yourself using hired equipment.

Preparing the wall for injection

If you want to carry out the work yourself, use a hired pressure-injection machine. You will need 68 to 90 litres (15 to 20 gallons) of DPC fluid for every 30m (100ft) of a wall 225mm (9in) thick.

Remove skirtings and hack off plaster and render to a height of 450mm (1ft 6in) above the line of visible damp. Repair and repoint the brickwork.

Drilling the injection holes

Drill a row of holes about 150mm (6in) above external ground level, but below a suspended wooden floor or just above one made of solid concrete. If the wall has an old DPC, set the new course just above it and take care not to puncture it when drilling. Use a masonry drill about 18 to 25mm (¾ to 1in) in diameter – but not smaller than the injecting nozzles of the machine. If possible, drill a row of identical holes from both sides of a wall 225mm (9in) or more thick to provide a continuous DPC.

When you are drilling a 225mm (9in) solid-brick wall, the holes should be at 112mm (4½in) centres, about 25mm (1in) below the upper-edge of a brick course. Angle them downwards slightly. Drill 75mm (3in) deep, unless the treatment is to be limited to one side only – in which case you should drill to a depth of 190mm (7½in). Treat each leaf of a cavity wall separately, drilling to a depth of 75mm (3in) in each leaf.

If the wall is made of impervious stone, you will need to drill into the mortar course around each stone block at the proposed DPC level, spacing the holes every 75mm (3in).

Injecting the fluid

Although there are various kinds of injection pump available, most of them work in basically the same way. With most types, the pump's filtered suction hose is inserted into a drum containing chemical. Make sure that the valves controlling the injection nozzles are closed, then connect the pump to the mains electrical supply.

Pressure injection machines usually have three to six nozzles. Connect the nozzles to the ends of the hoses and push them into the holes in the wall; if you are treating a thick wall, drill holes 75mm (3in) deep to begin with and start with the shorter nozzles.

Tighten the wing nuts sufficiently to secure the nozzles and form a seal – but don't overtighten them, or you may damage the expansion nipples at their tips. Open the control valves on all the nozzles except for the one at the far end, then switch on the pump so the fluid will circulate through the machine.

By opening its valve, bleed off some fluid through the remaining nozzle into a container to expel air from the system. Switch off the pump and insert the nozzle in the wall. Reopen the valve and allow the fluid to be injected until it wets the surface of the bricks. Maintain the pressure at about 100psi (pounds per square inch) by adjusting the valve on the pump body.

Close off all valves, then move the nozzles to the next series of holes and repeat the procedure. When you reach the other end of the wall, switch off the pump, then return to the starting point and drill the same holes to 190mm (7½in) deep. Swap the short injection nozzles for the longer 190mm (7½in) ones (you may need to wrap PTFE sealing tape round the threads), slot them into the wall, tighten their nuts, and inject the fluid. Flush the machine through with white spirit after use to clean out all traces of fluid.

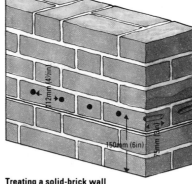

Treating a solid-brick wall

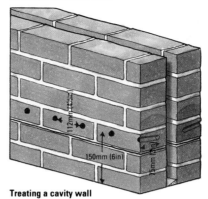

Treating a cavity wall

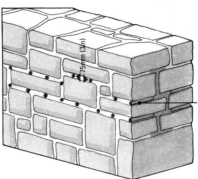

Treating an impervious-stone wall

● **Chemical-injection kit**
As an alternative to hiring equipment, you can buy a DIY kit that contains all the essential materials, tools and specialized equipment, together with full instructions.

At least one third thickness. Seek local professional advice.

Hiring equipment
Any tool-hire firm will supply you with all the materials and equipment necessary for injecting a chemical DPC yourself. It is an economical method that requires careful work rather than experience. Flush the machine thoroughly before you return it.

13

TREATING
A BASEMENT
OR CELLAR

Treating a wall with bitumen-latex emulsion
1 Skim coat of mortar
2 Coat of bitumen latex
3 Blinded coat of latex
4 Plaster or dry lining

Moisture-curing polyurethane
Damp-proof a floor with three or four coats of urethane applied with a broom.

14

Being at least partly below ground level, the walls and floors of a cellar or basement invariably suffer from damp to some extent. Because the problem cannot be tackled from the outside in the normal way, you will have to seal out the damp by treating the internal surfaces. Rising damp in concrete floors, whatever the situation, can be treated as described below, but penetrating or rising damp in walls other than in a cellar should be cured at source – since merely sealing the internal surfaces may encourage the damp eventually to penetrate elsewhere. In addition, ensure a treated cellar is properly ventilated and, if need be, even heat it to avoid condensation in the future.

Treating the floor

When laying a new concrete floor, incorporate a damp-proof membrane (DPM) during its construction. If the DPM was omitted or has failed in an existing floor, seal it with a heavy-duty, moisture-curing polyurethane.

Preparing the surface
Make sure that the floor is clean and grease-free. Fill any cracks and small holes by priming them with one coat of urethane, then one hour later applying a mortar made from 6 parts sand : 1 part cement, plus sufficient urethane to produce a stiff paste. Although it is possible to apply urethane to a damp or dry surface, it will penetrate a dry

floor better; so force-dry an excessively damp cellar with a fan heater before treatment. Remove all heaters from the room before you begin damp-proofing.

Applying urethane
Apply the first coat of urethane with a broom, using 1litre to cover about 5sq m (50sq ft). If you are damp-proofing a room with a DPC in the walls, take the urethane coating up behind the skirting to meet it. Two or three hours later, apply a second coat (further delay may result in poor inter-coat adhesion). Apply three or four coats in all. After three days, you can lay a conventional floorcovering or use the floor as it is.

Patching active leaks

Before you damp-proof a cellar, patch any cracks that are active water leaks, using a quick-drying hydraulic cement. Supplied in powder form, ready for mixing with water, the cement expands as it hardens, sealing out the moisture.

Undercut a crack or hole, using a chisel and club hammer. Mix some cement and hold it in a gloved hand until it is warm, then push it into the crack. Keep it in place with your hand or a trowel for three to five minutes, until it is hard.

TREATING THE WALLS

Moisture-cured polyurethane can be used to completely seal the walls of a cellar or basement, as well as the floor.

If you have decided to paint the walls, decorate with emulsion paints within 24 to 48 hours after treatment for maximum adhesion.

If you want to hang a wallcovering, apply two coats of emulsion first and use a heavy-duty paste. (Don't hang an impervious wallcovering, such as vinyl, as it's vital that the wall can breathe.)

Bitumen-latex emulsion
Where you plan to plaster or dry-line the basement walls, you can seal out the damp by using a relatively cheap bitumen-latex emulsion. However, it is not suitable as an unprotected covering, either for walls or floors, although it is often used as an integral DPM under the top screed of a concrete floor and as a waterproof adhesive for some tiles and for wooden parquet flooring.

Hack off old plaster to expose the brickwork, then apply a skim coat of mortar to smooth the surface. Paint the wall with two coats of the bitumen emulsion, joining with the DPM in the floor. Before the second coat dries, imbed some clean, dry sand (blinding) into it to provide a key for the coats of plaster (see below left).

Cement-based waterproof coating
In a cellar or basement where there is severe damp, apply a cement-based waterproof coating. Hack off old plaster or rendering in order to expose the wall.

To seal the join between a concrete floor and the wall, cut a chase about 20mm (¾in) wide by the same depth. Brush out the debris and fill the channel with hydraulic cement (see left), finishing it off neatly as an angled fillet.

Mix the powdered cement-based coating with an acrylic solution, following the manufacturer's instructions, then apply two coats to the wall with a bristle brush. However, when brick walls are damp, they bring salts to the surface in the form of white crystals known as efflorescence – so before treating with waterproof coating, apply a salt-inhibiting render made of 1 part sulphate-resisting cement : 2 parts clean rendering sand. Add 1 part liquid bonding agent : 3 parts of the mixing water. Apply a thin trowelled coat to a rough wall, or brush it onto a relatively smooth surface. Then leave it to set.

Rot occurs in unprotected household timbers, fences and outbuildings that are subject to damp. Fungal spores, which are always present, multiply and develop in these conditions until eventually the timber is destroyed. Fungal attack can cause serious structural damage and requires immediate attention if costly repairs are to be avoided. The two most common scourges are wet rot and dry rot.

Recognizing rot

Signs of fungal attack are easy enough to detect – but certain strains are much more damaging than others, and so it is important to be able to identify them.

Mould growth
White furry deposits or black spots on timber, plaster or wallpaper are mould growths. Usually, these are the result of condensation. When they are wiped or scraped off, the structure shows no sign of physical deterioration apart from staining. Cure the source of the damp conditions, and treat the affected area with a fungicide or a solution of 16 parts warm water : 1 part bleach.

Wet rot

Wet rot only occurs in timber that has a high moisture content. Once the cause of the moisture is eliminated, further deterioration is arrested. Wet rot often attacks the framework of doors and windows that have been neglected, allowing rainwater to penetrate joints or between brickwork and adjacent timbers. The first sign is often peeling paintwork. Stripping the paint reveals timber that is spongy when wet, but dark brown and crumbly when dry. In advanced stages the grain splits, and thin dark-brown fungal strands will be evident on the timber. Always treat wet rot as soon as practicable.

Dry rot

Once it has taken hold, dry rot is a most serious form of decay. Urgent treatment is essential. It will attack timber with a much lower moisture content than wet rot, but only in badly ventilated confined spaces indoors – unlike wet rot, which thrives outdoors as well as indoors .

Dry rot exhibits different characteristics depending on the extent of its development. It spreads by sending out fine pale-grey strands in all directions (even through masonry) to infect drier timbers and will even pump water from damp wood. The rot can progress at an alarming rate. In very damp conditions, these 'tubules' are accompanied by white growths resembling cotton wool. These are known as mycelium. Once established, dry rot develops wrinkled, pancake-shaped fruiting bodies, which produce rust-coloured spores that are expelled to rapidly cover surrounding timber and masonry. Infested timber becomes brown and brittle, with cracks across and along the grain, causing it to break up into cube-like pieces. You may detect a strong, musty, mushroom-like smell, produced by the fungus.

Wet rot – treat it at the earliest opportunity.

Dry rot – urgent treatment is essential.

TREATING ROT

Dealing with wet rot
Once you have eliminated the cause of the damp, cut away and replace badly damaged wood, then paint the new and surrounding woodwork with three liberal applications of chemical wet-rot eradicator. Brush the liquid well into the joints and end grain.

Before decorating, you can apply a wood hardener to reinforce slightly damaged timbers, then six hours later use wood filler to rebuild the surface. Repaint as normal.

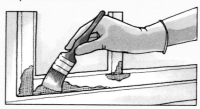

Paint rotted timbers with wood hardener

Dealing with dry rot
Dry rot requires more drastic action and should be treated by a specialist contractor unless the outbreak is minor and self-contained. The fungus is able to penetrate masonry, so look under the floorboards in adjacent rooms and check cavity walls for signs of rot.

Eliminate the source of dampness and ensure that there is adequate ventilation in roof spaces or under the floors by unblocking or installing air bricks. Cut out all infected timber up to at least 450mm (1ft 6in) beyond the last visible sign of rot. Chop plaster from nearby walls, following the strands. Continue for another 450mm (1ft 6in) beyond the extent of the growth. Collect all debris in plastic bags and burn it.

Use a chemical dry-rot eradicator to kill any remaining spores. Wire-brush the masonry, then apply three generous brushcoats to all timber, brickwork and plaster within 1.5m (5ft) of the infected area. An alternative method is to hire a coarse sprayer and go over the same area three times.

If a wall has been penetrated by strands of dry rot, drill regularly spaced, staggered holes into it from both sides. Angle the holes downwards, so the fluid will collect in them to saturate the wall internally. Patch holes after treatment.

Coat all replacement timbers with eradicator and immerse the end grain in a bucket of fluid for five to ten minutes. When you come to make good the wall, apply a zinc-oxychloride plaster.

15

PREVENTATIVE
TREATMENT

Since fungal attack can be so damaging, it is well worth taking precautions to prevent it occurring. Regularly decorate and maintain doorframes and window frames (where water can easily penetrate) and seal around them with mastic. Provide proper ventilation between floors and ceilings, and also in the loft. Check and eradicate sources of damp, such as plumbing leaks; and (the most important precaution of all, perhaps) apply a chemical preserver to unprotected timbers during routine maintenance.

Looking after timberwork

Treat new and existing timbers with a chemical preserver. Brush and spray two or three applications on standing timbers, paying particular attention to joints and end grain.

Immersing timbers

Timber in contact with the ground will benefit from prolonged immersion in preserver. Stand fence posts on end in a bucket of fluid for 10 minutes. For smaller timbers, make a shallow bath from loose bricks and line it with thick polyethylene sheet. Fill the trough with preserver and immerse the timbers, weighing them down with bricks to prevent them from floating (1). To empty the bath, sink a bucket at one end of the trough, then remove the bricks at that end so the fluid will pour out (2).

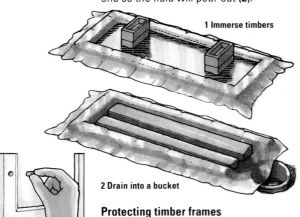

1 Immerse timbers

2 Drain into a bucket

Protecting timber frames

To protect timber frames, insert preserver in solid-tablet form into holes drilled at regular 50mm (2in) intervals in a staggered pattern. If the timber becomes wet, the tablets dissolve, placing preserver exactly where it is needed. Fill the holes with wood filler and paint as normal.

Protecting joints
Place preservative tablets close to the joints of a frame.

WOOD PRESERVERS

Water-based preservers are odourless and can safely be used on horticultural timbers. Most modern solvent-based products are also harmless to plants when dry, but it makes sense to check before you buy.

Clear preservatives

You can use clear liquid preservers that protect timber from dry or wet rot only. Alternatively, use an all-purpose fluid that will also provide protection against wood-boring insects. Clear preservers are useful when you want to retain the appearance of natural timber – oak beams or hardwood doors, for example – and you can usually paint or varnish the surface once the wood has dried.

Green preserver

There is also a green solvent-based preserver that is traditionally used for horticultural timbers. Being coloured, it helps to identify treated timbers for the future. However, the colour is due to the presence of copper, which is not a permanent colouring agent when used outdoors. Nevertheless, its protective properties are unaffected, even when the colour is washed out by heavy rain.

Wood-coloured preservers

Tinted preservers are formulated to protect sound exterior timbers against fungal and insect attack while staining the wood at the same time.

There is a choice of brown shades intended to simulate the most common hardwoods, and one that is designed specifically to preserve the richness of cedarwood. Solvent-based preservers are made with light-fast pigments that inhibit fading. They do not penetrate as well as a clear preserver, but generally offer slightly better protection than the coloured water-based preservers.

Clear   Coloured   Green

SAFETY WITH PRESERVERS

Solvent-based preservers are flammable, so do not smoke while using them and extinguish any naked lights. Wear protective gloves and goggles when applying preservers, together with a respirator when using these liquids indoors. Provide good ventilation while working, and do not sleep in a freshly treated room for 48 hours or so – in order to allow time for the fumes to dissipate completely. Wash spilt preserver from your skin and eyes with water immediately, and do not delay seeking medical advice if irritation persists.

Inevitably, old wooden casements and sash windows will have deteriorated to some extent, but regular maintenance and prompt repairs will preserve them almost indefinitely. New frames, or frames which have been stripped, should always be treated with a clear wood preserver before they are painted.

Regular maintenance

The bottom rail of a softwood sash is most vulnerable to rot, particularly if it is left unprotected. Rainwater seeps in behind old shrunken putty and moisture is gradually absorbed through cracked or flaking paintwork. Carry out an annual check and deal with any faults. Cut out old putty that has shrunk away from the glass and replace it. Remove flaking paint, make good any cracks in the wood with flexible filler and repaint. Do not forget to paint the underside of the sash.

Replacing a sash rail

Where rot is well advanced and the rail is beyond repair it should be cut out and replaced. This should be done before the rot spreads to the stiles, otherwise you will eventually have to replace the whole sash frame.

Remove the sash by unscrewing the hinges or, if it is a double-hung sash window, by removing the beading.

With a little care the repair can be carried out without removing the glass, though if the window is large it is safer to do so. In any event, cut away the putty from the damaged rail.

The bottom rail is tenoned into the stiles (1), but it can be replaced, using bridle joints. Saw down the shoulder lines of the tenon joints (2) from both faces of the frame and remove the rail.

Make a new rail, or buy a length of moulding if it is a standard section, then mark and cut it to length with a full-width tenon at each end. Set the positions of the tenons to line up with the mortises of the stiles. Cut the shoulders to match the rebated sections of the stiles (3) or, if there is a decorative moulding, pare the moulding from the stile to leave a flat shoulder (4). Cut slots in the ends of the stiles to receive the tenons.

Glue the new rail securely into place with a waterproof resin adhesive and reinforce the two joints with pairs of 6mm (¼in) stopped dowels. Drill the stopped holes from the inside of the frame and stagger them.

When the adhesive is dry, plane the surface as required and treat the new wood with a clear preserver. Reputty the glass and apply paint as soon as the putty is firm.

REPLACING A FIXED-LIGHT RAIL

The frames of some fixed lights (windows) are made like sashes, but are screwed permanently to the jamb and mullion. Such a frame can be repaired in the same way as a sash (see left) after its glass is removed and it is unscrewed from the window frame. Where this proves too difficult you will have to carry out the repair *in situ*.

First remove the putty and the glass, then saw through the rail at each end, close to the stile. Use a chisel to pare away what remains of the rail and chop out the tenons from the stiles. Cut a new length of rail to fit between the stiles and cut housings in its top edge at both ends to take loose tenons (1). Place the housings so that they line up with the mortises and make each housing twice as long as the depth of the mortise.

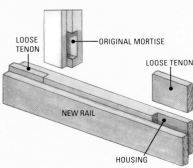

1 Cut housings at each end for loose tenons

Cut two loose tenons to fit the housings and two packing pieces. The latter should have one sloping edge (2).

Apply an exterior woodworking adhesive to all of the jointing surfaces, place the rail between the frame members, insert the loose tenons and push them sideways into the mortises. Drive the packing pieces behind the tenons to lock them in place. When the adhesive has set, trim the top edges, treat the new wood with clear preserver, replace the glass and reputty. Repaint once the putty is firm.

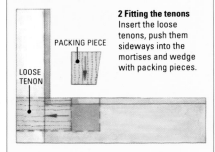

2 Fitting the tenons
Insert the loose tenons, push them sideways into the mortises and wedge with packing pieces.

● **Removing glass**
Removing glass from a window frame in one piece is not easy, so be prepared for it to break. Apply adhesive tape across the glass to bind the pieces together if it should break. Chisel away the putty to leave a clean rebate, then pull out the sprigs. Work the blade of a putty knife into the bedding joint on the inside of the frame to break the grip of the putty. Steady the glass and lift it out when it is freed.

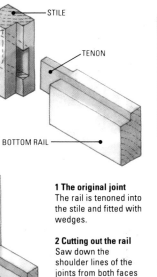

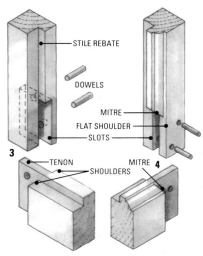

1 The original joint
The rail is tenoned into the stile and fitted with wedges.

2 Cutting out the rail
Saw down the shoulder lines of the joints from both faces of the frame.

3 Cutting the joint
Cut tenons at each end of the rail with the shoulders matching the sections of the stiles.

4 Moulded frames
Pare away the moulding of the stile to receive the square shoulder of the rail. Mitre the moulding.

REPAIRING
ROTTEN SILLS

The sill is a fundamental part of a window frame, and if one is afflicted by rot it can mean major repair work.

A casement-window frame is constructed in the same way as a doorframe and can be repaired in a similar way. All the glass should be removed first. The window board may also have to be removed, then refitted level with the replacement sill.

Make sure that the damp-proofing of the joint between the underside of the sill and the wall is maintained. Modern gun-applied mastics have made this particular problem easier to overcome. Some traditional frames have a galvanized-iron water bar between the sill and sub-sill. When replacing a sill of this type without removing the whole frame you may have to discard the bar and rely on mastic sealants to keep the water out.

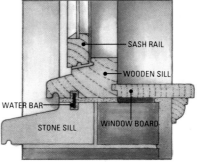

Traditional frame with stone sub-sill

Replacing a wooden sill

Do not simply replace a sill by cutting through it and fitting a new section between the jambs. Even if you seal the joints with mastic, any breakdown of the sealant will allow water to penetrate the brickwork and end grain of the wood, and you may find yourself doing the job all over again.

Serious rot in the sill of a sash window may require the whole frame to be taken out. Make and fit a new sill, using the old one as a pattern. Treat the new wood with a preserver and take the opportunity to treat the old wood which is normally hidden by the brickwork. Apply a bead of mastic sealant to the sill, then replace the complete frame in the opening from inside. Make good the damaged plaster.

It is possible to replace the sill from the inside with the frame in place (see right). Saw through the sill close to the jambs and remove the centre portion. Cut away the bottom ends of the inner lining level with the pulley stiles and remove the ends of the old sill. Cut the ends of the new sill to fit round the outer lining, and under the stiles and inner lining. Fit the sill and nail or screw the stiles to it.

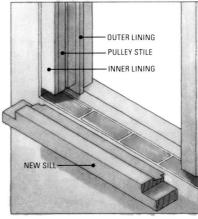

Cut the new sill to fit the frame

Repairing a stone sub-sill

The traditional stone sills that feature in older houses may become eroded by the weather if they are not protected with paint. They are also liable to crack if the wall subsides.

Repair cracks and eroded surfaces with a ready-mixed quick-setting waterproof mortar. Rake out the cracks to clean and enlarge them. Dampen the stone with clean water and work the mortar well into the cracks, finishing flush with the top surface.

Undercut any depressions caused by erosion to help the mortar adhere – a thin layer of mortar simply applied to a shallow depression in the surface will not last for long. Use a cold chisel to cut away the surface of the sill at least 25mm (1in) below the finished level and remove all traces of dust.

Make a wooden former to the shape of the sill and temporarily nail it to the brickwork. Dampen the stone, trowel in the mortar and tamp it level with the former, then smooth it out. Leave the mortar to set for a couple of days before removing the former. Allow it to dry thoroughly before applying paint.

Make a wooden former to the shape of the sill

CASTING A NEW SUB-SILL

Cut out what remains of the old stone sill with a hammer and cold chisel. Make a wooden mould with its end pieces shaped to the same section as the old sill. The open top of the mould represents the underside of the sill.

Fill two-thirds of the mould with fine-aggregate concrete, tamped down well. Add two lengths of mild-steel reinforcing rod, judiciously spaced to share the volume of the sill, then fill the remainder of the mould. Set a narrow piece of wood such as a dowel into notches cut in the ends of the mould. This is to form a 'throat' or drip groove in the underside of the sill.

Cover the concrete with polyethylene sheeting or dampen it regularly for two to three days to prevent rapid drying. When the concrete has set (allow about seven days), remove it from the mould and lay the new sill in the wall on a bed of mortar, packed from underneath with slate to meet the wooden sill.

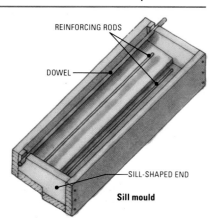

Sill mould

READY-MADE WINDOWS

Joinery suppliers offer a range of ready-made window frames in both hardwood and softwood. Some typical examples are shown below.

Manufactured wooden frames are treated with preserver and some are ready-primed for painting or prestained for final finishing. With the increased awareness of energy conservation, most frames are rebated to take double-glazed sealed units as well as traditional single glazing.

In addition to the stays and fasteners supplied with the frames, some windows also have the top rail of the opening sash, or the frame itself, slotted to take a ventilator kit to comply with Building Regulations for background ventilation in habitable rooms.

Self-assembly kits composed of machined framing are available to make frames of any size to fit non-standard window openings.

Casement windows

Vertical sliding sash windows

The style of the windows is important to the appearance of any house. If you are replacing windows in an older dwelling it is preferable – and not necessarily more expensive – to have new wooden frames made to measure rather than change to modern windows of aluminium or plastic.

Planning and Building Regulations

Window conversions do not normally need planning permission as they come under the heading of house improvement or home maintenance, but if you plan to alter your windows significantly – for example by bricking one up or making a new window opening, or both – you should consult your local Building Control Officer.

All authorities require minimum levels of ventilation to be provided in the habitable rooms of a house, and this normally means that the openable part of windows must have an area at least one-twentieth that of the room. Also, part if not all of a top vent must be 1.75m (5ft 9in) above the floor. Trickle ventilators with a $4000mm^2$ ($6\frac{1}{2}$ sq in) opening are also required for new installations.

If you live in a listed building or in a conservation area, you should also check with your local authority before making any changes to your windows.

Buying replacement windows

Try to maintain the character of an older house by preserving the original joinery. If you have to replace the window, copy the original style – specialist joinery firms will make up wooden frames to fit. Specify an appropriate hardwood or, for a painted finish, softwood impregnated with a timber preserver.

Alternatively, you can approach a replacement-window company, though this is likely to limit your choice to aluminium or plastic frames. Ready-glazed units can be fitted to your old timber sub-frames or to new hardwood ones supplied by the installer. Most replacement-window companies also fit the windows they supply, and their service includes disposing of the old windows and debris.

This method saves time and effort, but you should carefully consider the compatibility of such windows with the style of your house. Choose a frame that reproduces the proportions and method of opening of the original window as closely as possible.

Measure the width and height of the window opening. If the replacement window needs a timber sub-frame (and the existing one is in good condition), take your measurements from inside the frame. Otherwise, take them from the brickwork. You may have to cut away some of the rendering or internal plaster first in order to obtain accurate measurements. Order your replacement window accordingly.

Remove the old window by first taking out the sashes and then the panes of glass in any fixed light. Unscrew exposed fixings, such as may be found in a metal frame, or chisel away the plaster or rendering and cut through the fixings with a hacksaw. It should be possible to knock the frame out in one piece, but if not, saw through it in several places and lever the pieces out with a crowbar **(1)**. Clean up the exposed brickwork with a bolster chisel to make a neat opening.

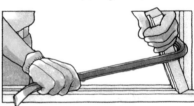

1 Lever out the pieces of the old frame

Cut the horns off the new frame, then wedge the frame in the window opening and check it is plumb **(2)**. Drill screw holes through the stiles into the brickwork **(3)**, then remove the frame and plug the holes or use frame fixings. Attach a bituminous-felt damp-proof course to the jambs and sill and refit the frame, checking again that it is plumb before screwing it firmly in place.

Make good the wall with mortar and plaster. Gaps of 6mm ($\frac{1}{4}$in) or less can be filled with mastic. Glaze the new frame as required.

2 Fit the new frame   **3 Drill fixing holes**

REPLACEMENT WINDOWS

Joining frames
A flexible sealant is used for joining standard frames. The frames are screwed together to fit the opening.

Bay windows

A bay window is a combination of window frames built out from the face of the building. The side frames may be set at 90- or 45-degree angles to the front of the house. Curved bays are also made with equal-sized frames set at a very slight angle to each other to form a faceted curve.

The brick structure that supports the window frames may continue up through all storeys, finishing with a gable roof. Alternatively, the bay might have a brick base only, or be supported on brackets, with a flat or pitched roof.

Bay windows can break away from the main wall as a result of subsidence caused by poor foundations or differential ground movements. Damage from slight movement can be repaired once it has stabilized by repointing the brickwork and applying mastic sealant to gaps round the woodwork. However, any damage from extensive or persistent movement should be dealt with by a builder. Consult your local Building Control Officer and inform your insurance company.

Fitting the frame
Where the height of the original window permits, fit standard window frames to make up a replacement window. Various combinations of frames can be joined with shaped hardwood corner posts to set the side frames at an angle of 90 or 45 degrees. A sealant is used to weatherproof the joints between the posts and frames.

90-degree-angle bay

45-degree-angle bay

STANDARD FRAMES

POST
SILL

Standard frames joined with posts

Modern angled bay with decorative lead flashing

Bow windows

These windows are constructed on a shallow curve and normally project from a flat wall. Complete hardwood bow-window frames are available from joinery suppliers, ready for installation in a brickwork opening. A flat-topped canopy of moulded plastic is made for finishing the top of the window in place of a traditional lead-sheet covering.

Fitting the frame
Tack damp-proof-course material to the sides of the frame and the underside of the sill, then fit both the frame and the canopy into the wall opening, the outer edges of the frame set flush with the wall. Screw the frame to the brickwork. The vertical damp proofing should overlap any damp-proof course built into the wall.

Weatherproof the canopy with a lead flashing cut into the wall and dressed over the upturned rear edge of the canopy. Use mastic to seal the joints between the frame and the brickwork.

Attractive bow window that suits an older house

REPLACING A SASH WINDOW

Traditional boxed-sash windows fitted with cords and counterweights can be home-made or supplied by specialists who can also fit them for you. Alternatively, an old vertically sliding sash window can be replaced with a new frame with spiral-balance sashes.

Remove the sashes, then take out the old frame from inside the room. Prise off the architrave, then the window boards, and chop away the plaster as necessary. Most frames are wedged in their openings, and you can loosen one by simply hitting the sill on the outside with a heavy hammer and a wood block. Lift out the frame **(1)** and remove any debris from the opening.

Fit a traditional sash-window frame exactly like the original, making sure the wood is treated with preserver.

Set a new spring-balance type (which has a thinner frame) centrally in the window opening. Check the frame for plumb and wedge the corners at the head and sill. Make up the space left by the old box stiles with mortared brickwork **(2)**.

Metal brackets screwed to the new frame's jambs can also be set in the mortar joints to secure the frame.

When the mortar has set, replaster the interior wall and replace the architrave. Glaze the sashes and apply a mastic sealant to the joints between the exterior brickwork and the frame to keep rainwater out.

1 Lift out old frame   **2 Fill gaps with brick**

TYPES OF GLASS

Glass is made from silica sand, which is heated with such additives as soda, lime and magnesia until it is molten in order to produce the raw material. The type and quality of glass produced for windows is determined by the method by which it is processed at the molten stage. Ordinary window glass is known as annealed glass. Special treatments during manufacture give glass particular properties, such as heat-resistance or extra strength.

Float glass

Float glass is generally used for glazing windows. It is made by floating the molten glass on a bath of liquid tin to produce a sheet with flat, parallel and distortion-free surfaces. It has virtually replaced plate glass, which was a rolled glass ground and polished on both sides.

Clear float glass is manufactured in a range of thicknesses from 3mm (1/8in) up to 25mm (1in), but it is generally stocked in only three: 3mm (1/8in), 4mm (5/32in) and 6mm (1/4in).

Patterned glass

Patterned glass has one surface embossed with a texture or a decorative design. It is available in clear or tinted sheets in thicknesses of 3mm (1/8in), 4mm (5/32in), and 6mm (1/4in).

The transparency of the glass depends to a large extent on the density of the patterning. Such 'obscured' glass is used where maximum light is required while maintaining privacy, such as for bathroom windows. Patterned safety glass – toughened or laminated – should be used for bath or shower screens.

Solar-control glass

Special glass which reduces the heat of the sun is often used for roof-lights. This tinted glass, which can be of the float, laminated or textured type, also reduces glare, though at the expense of some illumination. Solar-control glass is available in thicknesses ranging from 4mm (5/32in) to 12mm (1/2in), depending on the type. The 6mm (1/4in) thick glass is the one most commonly used.

Low-emissivity glass

Low-E glass is a clear float glass with a special coating on one surface. It is primarily used for the inner pane of double glazing with the coating facing the cavity to produce an efficient unit that retains warmth in the room while keeping the cold out. It provides good light transmission and optimizes the heat from the sun. The outer pane of the unit can be of any other type of glass.

Non-reflective glass

This type of glass is 2mm (1/16in) thick with a slightly textured surface, and is much used for glazing picture frames. When placed within 12mm (1/2in) of the picture surface, the glass appears completely transparent yet eliminates the surface reflections associated with ordinary polished glass.

Safety glass

Glass which has been strengthened by means of reinforcement or a toughening process is known as safety glass. It should be employed whenever the glazed area is relatively large or where its position makes it vulnerable to accidental breakage. In domestic situations, safety glass should be considered for glazed doors, low-level windows and shower screens.

Wired glass

Wired glass is a 6mm (1/4in) thick roughcast or clear annealed glass with a fine steel wire mesh incorporated in it during its manufacture. Glass with a 13mm (1/2in) square mesh is known as Georgian wired glass. The mesh stops the glass disintegrating in the event of breakage. The glass itself is not special and is no stronger than ordinary glass of the same thickness.

Wired glass is regarded as a fire-resistant material with one-hour rating. Though the glass may break, its wire reinforcement helps to maintain its integrity and prevent the spread of fire.

Toughened glass

Toughened glass is ordinary glass that has been heat-treated to improve its strength. It is sometimes referred to as tempered glass. The process of treatment renders the glass about four to five times stronger than an untreated glass of similar thickness. In the event of its breaking it merely shatters into relatively harmless granules.

Toughened glass cannot be cut. Any work required, such as holes drilled for screws, must be done before the toughening process. Joinery suppliers of doors and windows usually stock standard sizes of toughened glass to fit standard frames.

Laminated glass

Laminated glass is made by bonding together two or more layers of glass with a clear tear-resistant plastic film sandwiched between. This glass absorbs the energy of an object hitting it, so preventing the object penetrating the pane. The plastic interlayer also binds broken glass fragments together and reduces the risk of injury from fragments of flying glass.

It is made in a range of thicknesses from 4mm (5/32in) up to 8mm (5/16in), depending on the type of glass used. Clear, tinted and patterned versions are all available.

Etched glass

Following the increased interest in the renovation of older houses, traditional acid-etched decorative glass is now available. It is produced in 4mm (5/32in) and 6mm (1/4in) thicknesses and in four designs. It can be toughened or laminated for safety.

Patterned and tinted glass
Clear and tinted glass can be used for restoring windows in older houses as well as new installations.

WORKING
WITH GLASS

BUYING GLASS

You can buy most types of glass from your local stockists. They will advise you on thickness, cut the glass to your measurements and deliver larger sizes and amounts.

The thickness of glass, once expressed by weight, is now measured in millimetres. If you are replacing old glass, measure its thickness to the nearest millimetre, and, if it is slightly less than any available size, buy the next one up for the sake of safety.

Though there are no regulations about the thickness of glass, for safety reasons you should comply with the recommendations set out in the British Standard Code of Practice. The required thickness of glass depends on the area of the pane, its exposure to wind pressure and the vulnerability of its situation – for example, in a window overlooking a play area. Tell your supplier what the glass is needed for – a door, a window, a shower screen – to ensure that you get the right type.

Measuring up

Measure the height and width of the opening to the inside of the frame rebate, taking the measurement from two points for each dimension. Also check that the diagonals are the same length. If they differ markedly, indicating that the frame is out of square, make a cardboard template of it. In any case, deduct 3mm (⅛in) from the height and width to allow a fitting tolerance. When making a template, allow for the thickness of the glass cutter.

When you order patterned glass, specify the height before the width. This will ensure that the glass is cut with the pattern running in the right direction. Alternatively, take a piece of the old glass with you (which you may need to do in any case to match the pattern).

For an asymmetrically shaped pane of patterned glass supply a template, marking the surface that represents the outside face. This ensures that the glass will be cut with its smooth surface to the outside and will therefore be easier to keep clean.

Always carry panes of glass on edge to stop them bending, and use stout work gloves to protect your hands from sharp edges. Wear the same gloves and protect your eyes with goggles when removing broken glass from a frame. Wrap it up in thick layers of newspaper if you have to dispose of it in your dustbin, but check first with your local glazier to see if he might be willing to add it to his offcuts, which are usually sent back to the manufacturers for recycling.

Basic glass-cutting

It is usually unnecessary to cut glass at home as suppliers are willing to do it, but you may have some surplus glass that you wish to cut yourself. Diamond-tipped cutters are available, but the type with a steel wheel is cheaper and quite adequate for normal use.

Cutting glass successfully is largely a matter of practice and confidence. If you have not done it before, you should make a few practice cuts on waste pieces of glass and get used to the 'feel' before doing a real job.

Lay the glass on a flat surface covered with a blanket. (Patterned glass is placed patterned side downwards and cut on its smooth side.) Clean the surface with methylated spirit.

Set a T-square the required distance from one edge, using a steel measuring tape (**1**). If you are working on a small piece of glass or do not have a T-square, mark the glass on opposing edges with a felt-tipped pen or wax pencil and use a straightedge to join up the marks and guide the cutter.

Lubricate the steel wheel of the glass cutter by dipping it in thin oil or paraffin. Hold the cutter between middle finger and forefinger (**2**) and draw it along the guide in one continuous stroke. Use even pressure throughout and run the cut off the end. Slide the glass forward over the edge of the table (**3**) and tap the underside of the scored line with the back of the cutter to initiate the cut. Grip the glass on each side of the score line with gloved hands (**4**), lift the glass and snap it in two. Alternatively, place a pencil under each end of the scored line and apply even pressure on both sides until the glass snaps.

1 Measure the glass with a tape and T-square

2 Cut glass with one continuous stroke

3 Tap the underside to initiate the cut

4 Snap the glass in two

Cutting a thin strip of glass

A pane of glass may be slightly oversize due to inaccurate measuring or cutting or to distortion of the frame.

Remove a very thin strip of glass with the aid of a pair of pliers. Nibble away the edge by gripping the waste with the tip of the jaws close to the scored line.

Nibble away a thin strip with pliers

SEE ALSO
Details for:
Removing glass 24

Fitting items such as an extractor fan may involve cutting a circular hole in a pane of glass. This can be done with a beam-compass glass cutter.

Cutting a circle in glass

Locate the suction pad of the central pivot on the glass, set the cutting head at the required distance from it and score the circle round the pivot with even pressure. Now score another, smaller, circle inside the first one (1). Remove the cutter and score across the inner circle with straight cuts, then make radial cuts about 25mm (1in) apart in the outer rim. Tap the centre of the scored area from underneath to open up the cuts (2) and remove the inner area. Next tap the outer rim and nibble away the waste with pliers if necessary.

To cut a disc of glass, scribe a circle with the beam-compass cutter, then score tangential lines from the circle to the edges of the glass (3). Tap the underside of each cut, starting close to the edge of the glass.

1 Score the circle with even pressure

Smoothing the edges of cut glass

You can grind down the cut edges of glass to a smooth finish using wet-and-dry paper wrapped round a wooden block. It is fairly slow work, though just how slow will depend on the degree of finish you require.

Start off with medium-grit paper wrapped tightly round the block. Dip the block, complete with paper, in water and begin by removing each 'arris' (the sharp corners along the edge) with the block, holding it at 45 degrees to the edge. Keep the abrasive paper wet.

Follow this by rubbing down the actual edge to remove any nibs and go on to smooth it to a uniform finish. Repeat the process with progressively finer grit papers. Finally, polish the edge with a wet wooden block coated with pumice powder.

2 Tap the centre of the scored area

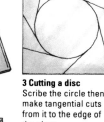
3 Cutting a disc
Scribe the circle then make tangential cuts from it to the edge of the glass.

Using a glass-cutting template

Semi-circular windows and glazed openings above Georgian-style doors have segments of glass mounted between radiating glazing bars.

Modern windows with semi-circular openings and reproductions of period doors can be glazed with ready-shaped panes available from joinery suppliers, but it is necessary to cut panes of glass to fit an old glazed door or window.

The pieces of glass are segments of a large circle, beyond the scope of a standard beam-compass glass cutter (see above), so make a cardboard template to serve as a guide for scoring with an ordinary cutter.

Remove the broken glass, clean up the rebate, then tape a sheet of paper over the opening and, using a wax crayon, take a rubbing of the shape (1). Remove the paper pattern and tape it to a sheet of thick cardboard. Make the actual template about 3mm (⅛in) smaller all round than the pattern to provide a tolerance for fitting between glazing bars, and to allow for the thickness of the glass cutter.

Fix the template to the glass with double-sided tape, score round it with the glass cutter (2), running all cuts to the edge, and snap the glass in the normal way. Any slight irregularities will be hidden by the glazing-bar rebates and putty.

1 Take a rubbing of the shape with a crayon

2 Cut round the template, using even pressure

● **Plastic glazing**
As an alternative to glass, cut acrylic plastic with a fret saw to fit awkward shapes.

Drilling a hole in glass

There are special spear-point drilling bits available for drilling holes in glass. As glass should not be drilled at high speed, use a hand-held brace or power drill set to run at a low speed.

Mark the position for the hole, no closer than 25mm (1in) to the edge of the glass, using a felt-tipped pen or a wax pencil. On mirror glass work from the back (the coated surface).

Place the tip of the bit on the marked centre and, with light pressure, twist it back and forth so that it grinds a small pit and no longer slides off the centre. Form a small ring with putty round the pit and fill the inner well with a lubricant such as white spirit, paraffin or water.

Work the drill at a steady speed and pressure – too much pressure may chip the glass. When the tip of the drill just breaks through, turn the glass over and drill from the other side. Drilling straight through from one side risks breaking out the surface round the hole.

Drilling glass
Always run the drill in a lubricant to reduce friction.

REPAIRING A BROKEN WINDOW

SEE ALSO
Details for:
Buying glass   22

A cracked window pane, even when no glass is missing, is a safety hazard. Smashed panes are a security risk and are no longer weatherproof, so replace them promptly.

Temporary repairs

For temporary protection from the weather a sheet of polyethylene can be taped or pinned with battens over the outside of the window frame. If the window is merely cracked it can be temporarily repaired with a special clear self-adhesive waterproof tape. Applied to the outside, the tape gives an almost invisible repair.

Safety with glass

The method you use to remove the glass from a broken window will to some extent depend on conditions. If the window is not at ground level, it may be safest to take out the complete sash to do the job. However, a fixed window will have to be repaired on the spot, wherever it is.

Large pieces of glass should be handled by two people and the work done from a tower rather than ladders. Avoid working in windy weather and always wear protective gloves and spectacles for this work.

Repairing glass in wooden frames

In wooden window frames the glass is set into a rebate cut in the frame's moulding and bedded in putty. Small wedge-shaped nails known as sprigs are also employed to hold the glass in place. Linseed-oil putty is used for glazing traditional softwood frames. A more flexible putty compound is made to accommodate movement in modern softwood or hardwood frames finished with paints or stains that are moisture-vapour permeable.

Removing the glass

If the glass in a window pane has shattered, leaving jagged pieces set in the putty, grip each piece separately (wearing gloves) and try to work it loose **(1)**. It is always safest to start working from the top of the frame.

Old putty that is dry will usually give way, but if it is strong it will have to be cut away with a glazier's hacking knife and a hammer **(2)**. Alternatively, use a blunt wood chisel. Work along the rebate to remove the putty and glass. Pull out the sprigs with pincers **(3)**.

If the glass is cracked but not holed, run a glass cutter round the perimeter of the pane about 25mm (1in) from the frame, scoring the glass **(4)**. Fasten strips of self-adhesive tape across the cracks and scored lines, then tap each piece of glass until it breaks free and is held only by the tape **(5)**. Carefully remove individual pieces of glass, working from the centre of the pane.

Clean remnants of old putty out of the rebates, then seal the wood with wood primer. Measure the height and width of the opening to the inside of the rebates and have your new glass cut 3mm (⅛in) smaller on each dimension to provide a tolerance for fitting.

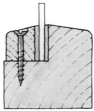

Glass fixed with putty

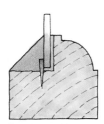

Wooden bead fixing
Some wooden frames feature screwed-on beading bedded into mastic to hold the panes in place. Unscrew beading and scrape out mastic. Bed new glass in fresh mastic and replace beading.

1 Work loose the broken glass

2 Cut away the old putty

3 Pull out the old sprigs

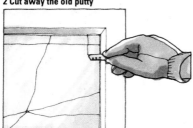

4 Score glass before removing a cracked pane

5 Tap the glass to break it free

Fitting new glass

Purchase new sprigs and enough putty for the frame. Your glass supplier should be able to advise you on this but, as a guide, 500g (1lb) of putty will fill an average-sized rebate of about 4m (13ft) in length.

Knead a palm-sized ball of putty to an even consistency. Very sticky linseed-oil putty is difficult to work with, so wrap it briefly in newspaper to absorb some of the oil. You can soften putty that is too stiff by adding linseed oil.

Press a fairly thin, continuous band of putty into the rebate all round with your thumb. This is the bedding putty. Lower the edge of the new pane on to the bottom rebate, then press it into the putty. Press close to the edges only, squeezing the putty to leave a bed about 2mm (¹⁄₁₆in) on the inside, then secure the glass with sprigs about 200mm (8in) apart. Tap them into the frame with the edge of a firmer chisel so that they lie flat with the surface of the glass **(1)**. Trim the surplus putty from the back of the glass with a putty knife.

Apply more putty to each rebate on the outside. With a putty knife **(2)**, work the putty to a smooth finish at an angle of 45 degrees. Wet the knife with water to prevent it dragging and make neat mitres in the putty at the corners. Let the putty set and stiffen for about three weeks, then apply a paint or stain finish as required. Before painting, clean any putty smears from the glass with methylated spirit. Let the paint lap the glass slightly to form a weather seal.

A self-adhesive plastic foam can be used instead of the bedding putty. Run it round the back of the rebate in a continuous strip, starting from a top corner, and press the glass into place on the foam. Secure it with sprigs, then apply putty on the outside as described above. Alternatively, apply a second strip of foam round the outside of the glass and cover it with a wooden beading, then paint.

1 Tap in new sprigs

2 Shape the putty

The great majority of external doorframes are constructed of softwood, and this, if it is regularly maintained with a good paint system, will give years of excellent service. However, the ends of door sills and the frame posts are vulnerable to wet rot if they are subject to continual wetting. This can happen when the frame has moved because the timber has shrunk, or where old pointing has fallen out and left a gap where water can penetrate. Alternatively, old and porous brickwork or an ineffective damp-proof course can be the cause of wet-rot damage.

Prevention is always better than cure, so check round the frame for any gaps and apply a mastic sealant where necessary. Keep all pointing in good order. A minor outbreak of wet rot can be treated with the aid of a proprietary repair kit and a chemical preserver.

It is possible for the sill to rot without the doorposts being affected, in which case just replace the sill. But if the posts are also affected, repair them at the same time (see right). In some cases the post ends are tenoned into the sill and fitted as a unit.

Replacing a sill

You can buy 150 x 50mm (6 x 2in) softwood or hardwood door-sill sections which can be cut to the required length. If your sill is not of a standard-shaped section, you can have a replacement made to order. A hardwood such as oak will be relatively expensive, but will prove more economical in the long run as it lasts much longer.

Taking out the old sill
First measure and note down the width of the door opening, then remove the door. The posts are usually tenoned into the sill, so split the sill lengthways with a wood chisel in order to dismantle the joints. A saw-cut across the centre of the sill makes the job easier.

The ends of the sill are set into the brickwork on each side of the opening. To release the sill, use a plugging chisel to chop out the mortar joints carefully, then pull out a brick from each side. Keep them for replacing later.

The new sill has to be inserted from the front so that it can be tucked under the posts and into the brickwork. Cut off the tenons level with the shoulders of the posts (1). Mark and cut shallow housings for the ends of the posts in the top of the new sill, spacing them apart as previously noted. The housings must be deep enough to take the full width of the posts (2) which may mean the sill being slightly higher than the original one, so you will have to trim a little off the bottom of the door.

Fitting a new sill
Try the new sill for fit and check that it is level. Before fixing it, apply two coats of all-purpose wood preserver to its underside and to both ends, and, as a precaution against rising or penetrating damp, apply two or three coats of bitumen latex emulsion to the brickwork in contact with the sill.

When both treatments are dry, glue the sill to the posts, using an exterior-grade woodworking adhesive. Wedge the underside of the sill with pieces of roofing slate to push it up against the ends of the doorposts. Skew-nail the posts to the sill and leave it for the adhesive to set.

Pack the gap between the underside of the sill and the masonry with a stiff mortar of 3 parts sand : 1 part cement, then rebond and point the bricks. Finish by treating the wood with a preserver and seal any gaps around the doorframe with mastic.

1 Cut tenons off level with the joint's shoulder

2 Cut a housing to receive the post

REPAIRING DOORPOSTS

Rot can attack the ends of doorposts where they meet stone steps or are set into concrete, especially in a doorway that is regularly exposed to driving rain.

If the damage is not too extensive the rotten end can be cut away and replaced with a new piece, either scarf-jointed or halving-jointed into place. If your sill is made from wood, combine the following information with that given for replacing a sill (see left).

First remove the door, then saw off the end of the affected post back to sound timber. For a scarf joint make the cut at 45 degrees to the face of the post (1); for a halving joint cut it square. If the post is located on a metal dowel set into the step, chop out the dowel with a cold chisel.

Measure and cut a matching section of post to length, allowing for the overlap of the joint, then cut the end to 45 degrees or mark and cut both parts of the post to form a halving joint (2).

Drill a hole in the end of the new section for the metal dowel if it is still usable. If not, make a new one from a piece of galvanized-steel gas pipe and prime it to prevent corrosion. Treat the new wood with a preserver and insert the dowel. Set the dowel in mortar and glue and screw the joint (3).

If a dowel is not used, fix the post to the wall with counterbored screws. Place hardboard or plywood packing behind it if necessary and plug the screw holes.

Apply a mastic sealant to the joints between the door post, wall and base.

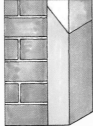

1 Scarf joint

2 Halving joint

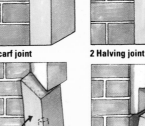

3 Set dowel in mortar as you close up either joint

FITTING AND HANGING DOORS

SEE ALSO
Details for:
Draught excluders   6

Whatever the style of door you wish to fit the procedure is similar, with only minor differences to contend with. Two good-quality 100mm (4in) butt hinges are enough to support a standard door, but if you are hanging a fire door or a heavy hardwood one you should add a third, central hinge.

As you will have to try a door in its frame several times to get the perfect fit, it pays to have someone working with you.

Fitting a door

Before attaching the hinges to a new door make sure that it fits nicely into its frame. It should have a clearance of 2mm (¹⁄₁₆in) at the top and sides and should clear the floor by at least 6mm (¼in). As much as 12mm (½in) may be required for a carpeted floor.

Measure the height and width of the door opening and the depth of the rebate in the frame into which the door must fit. Choose a door of the right thickness and, if you cannot get one that fits the opening exactly, select one large enough to be trimmed down.

Cutting to size
New doors are often supplied with 'horns', extensions to their stiles which prevent the corners being damaged while the doors are in storage. Cut these off with a saw (1) before starting to trim the door to size.

Transfer the measurements from the opening to the door, making allowance for necessary clearances all round.

To reduce the width of the door support it on edge, latch-stile up, in a portable bench, then plane the stile down to the marked line. If a lot of wood has to be removed, take some off each stile– this is especially important in the case of panel doors in order to preserve their symmetry.

If you need to take off more than 6mm (¼in) to reduce the height of the door, remove it with a saw and finish off with a plane. Otherwise plane the waste off (2). The plane must be extremely sharp to deal with the end grain of the stiles. Work from each corner towards the centre of the bottom rail to avoid 'chipping out' the corners.

Try the door in the frame, supporting it on shallow wedges (3). If it still does not fit take it down and remove more wood where appropriate.

1 Saw off horns

2 Plane to size

3 Wedge the door

Fitting hinges

The upper hinge is set about 175mm (7in) from the door's top edge and the lower one about 250mm (10in) from the bottom. They are cut equally into the stile and doorframe. Wedge the door in its opening and, with the wedges tapped in to raise it to the right floor clearance, mark the positions of the hinges on both the door and frame.

Stand the door on edge, hinge stile uppermost, open a hinge and, with its knuckle projecting from the edge of the door, align it with the marks and draw round the flap with a pencil (1). Set a marking gauge to the thickness of the flap and mark the depth of the housing. With a chisel, make a series of shallow cuts across the grain (2) and pare out the waste to the scored line. Repeat the procedure with the second hinge, then, using the flaps as guides, drill pilot holes for the screws and fix both hinges into their housings.

Wedge the door in the open position, aligning the free hinge flaps with the marks on the doorframe. Make sure that the knuckles of the hinges are parallel with the frame, then trace the housings on the frame (3) and cut them out as you did the others.

Adjusting and aligning
Hang the door with one screw holding each hinge and see if it closes smoothly. If the latch stile rubs on the frame you may have to make one or both housings slightly deeper. If the door appears to strain against the hinges it is said to be 'hinge bound'. In this case insert thin cardboard beneath the hinge flaps to pack them out. When the door finally opens and closes properly drive in the rest of the screws.

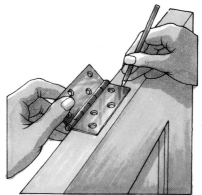
1 Mark round the flap with a pencil

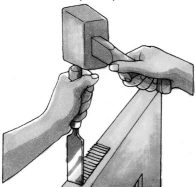
2 Cut across the grain with a chisel

3 Mark the size of the flap on the frame

MEASUREMENTS

A door that fits well will open and close freely and look symmetrical in the frame. Use the figures given as a guide for trimming the door and setting out the position of the hinges.

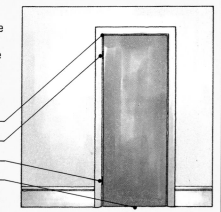

2mm (¹⁄₁₆) clearance at top and sides

Upper hinge 175mm (7in) from the top

Lower hinge 250mm (10in) from the bottom

6 to 12mm (¼ to ½in) gap at the bottom

Rising butt hinges

Rising butt hinges lift a door as it is opened and are fitted to prevent it dragging on thick pile carpet.

They are made in two parts: a flap with a fixed pin which is screwed to the doorframe, and another flap with a single knuckle which is fixed to the door. The knuckle pivots on the pin.

Rising butt hinges must be fixed one way up only, and are therefore made specifically for left-hand or right-hand opening. The countersunk screwholes in the fixed-pin flap indicate the side to which it is made to be fitted.

Fitting

Trim the door and mark the hinge positions (see opposite), but before fitting the hinges plane a shallow bevel at the top outer corner of the hinge stile so that it will clear the frame as it opens. As the stile runs through to the top of the door, plane from the outer corner towards the centre to avoid splitting the wood. The top strip of the doorstop will mask the bevel when the door is closed.

Fit the hinges to the door and frame, then lower the door on to the hinge pins, taking care not to damage the architrave above the opening.

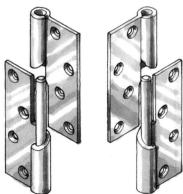

Left-hand opening   Right-hand opening

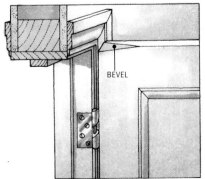

Plane a shallow bevel to clear the doorframe

Weatherproofing a door

Fitting a weatherboard

A weatherboard is a special moulding which is fitted to the bottom of an external door to shed rainwater away from the threshold. To fit one, measure the width of the opening between the doorstops and cut the moulding to fit, cutting one end at a slight angle where it meets the doorframe on the latch side. This will allow it to clear the frame as the door swings open.

Use screws and a waterproof glue to attach a weatherboard to an unpainted door. When fitting one to a door that is already finished, apply a thick coat of primer to the back surface of the weatherboard to make a weatherproof seal, then screw the moulding in place while the primer is still wet. Fill or plug the screw holes before you prime and finish the weatherboard.

Allowing for a weather bar

Though a rebate cut into the head and side posts of an external doorframe provides a seal round an inward-opening door, a rebate cut into the sill at the foot of the door would merely encourage water to flow into the house.

Unless protected by a porch, a door in an exposed position needs to be fitted with a weather bar to prevent rainwater running underneath. This is a metal or plastic strip which is set into the step or sill. If you are putting in a new door and wish to fit a weather bar, use a router or power saw to cut a rebate across the bottom of the door in order to clear the bar.

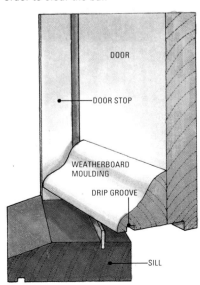

Door fitted with a weatherboard

ADJUSTING BUTT HINGES

Perhaps you have a door catching on a bump in the floor as it opens. You can, of course, fit rising butt hinges, but the problem can be overcome by resetting the lower hinge so that its knuckle projects slightly more than the top one. The door will still hang vertically when closed, but as it opens the out-of-line pins will throw it upwards so that the bottom edge will clear the bump.

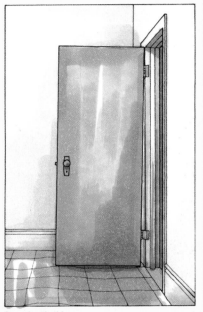

Resetting the hinge
You may have to reset both hinges to the new angle to prevent binding.

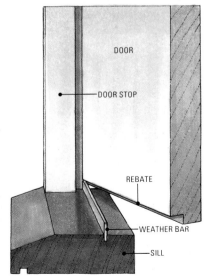

Sill fitted with weather bar

REPAIRS AND IMPROVEMENTS

SEE ALSO
Details for:
Wood preserver   16

Repairing a battened door

The battens, or tongue-and-groove boards, of a ledged and braced door tend to rot first along the bottom edge of the door because the end grain absorbs moisture. Nailing a board across the bottom of the door is not the easy solution it may appear because moisture will be trapped behind the board and will increase the rot.

Remove the door and cut back the damaged boards to sound material. Where a batten falls on a rail, use the tip of a tenon saw, held at a shallow angle, to cut through most of it, then finish off the cut with a chisel. Use a padsaw or a power jigsaw where the blade can pass clear of the rail. When replacing the end of a single batten make the cut at right angles (**1**). When a group of battens is to be replaced, make 45-degree cuts across them (**2**). In this manner the interlocking of the tongued and grooved edges between the old and new sections is better maintained.

When cutting new pieces of boarding to fit, leave them over length. Apply an exterior woodworking adhesive to the butting ends of the battens, but take care not to get any on the tongue-and-groove joints. Tap the pieces into place and nail each to the rail with two, staggered lost-head nails. Cut off the ends of the repaired battens in line with the door's bottom edge, then treat the wood with a preserver to prevent any further damage.

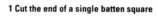

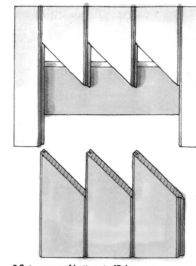

1 Cut the end of a single batten square

2 Cut a group of battens to 45 degrees

Easing a sticking door

If the bottom corner of a door rubs against the frame, take it off its hinges and shave the corner with a plane. If the top corner is rubbing check the hinges before planing. After years of use, hinges wear and the pins become slack, allowing the door to drop. In this case fit new hinges or, as a cheaper alternative, swap the old hinges, top for bottom, which reverses the wear on the pins.

Swopping hinges
Swop worn hinges, top for bottom, for a cheap and convenient repair.

REPAIRING A PANEL DOOR

A panel door is commonly used for the main entrance to a house. Unless subjected to serious neglect this type of door should give good service over the life of the building. However, even sound doors can be seriously damaged when a housebreaker uses brute force to gain entry. Although relatively strong, practically any entrance door can be kicked in or smashed open with a sledgehammer, the weakest point often being down the hinged edge rather than on the well-fortified lock side.

When the frame or panels are badly splintered the easiest course is to replace the whole door. However, if the door is unique and therefore worth preserving, insert pieces of new wood to repair the damage.

Rebuilding the edge
If the hinge stile has been split, the wood will have failed in the vicinity of the hinge screws and broken out from the front face of the door. If the splintered wood can be clamped back into place, glue the break with exterior wood adhesive. Cover the repair with a piece of polyethylene sheeting and place wooden blocks under the cramp heads to spread the forces over the damaged area. You will also need to glue wooden plugs into the old screw holes for refixing the door. Clean up the repair with a plane and fill any hollows with a wood filler prior to repainting.

However, it is likely that the split wood is beyond repair, in which case replace the damaged material with new wood. Use a chisel to cut back the damaged stile to sound wood, forming a regular recess. Undercut the ends to 45 degrees (see below).

Shape a block of similar wood to fit the length of the recess, but leave it oversize in width and thickness. Glue it in place and, when set, plane the block flush. Cut the housings for the hinges.

Rehang the door and fit hinge bolts to the door stile and frame to help prevent it happening again.

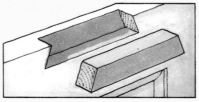

Glue a shaped block into the recess

Concrete is used in and around the house as a surface for solid floors, drives, paths and walls. In common with other building materials, it suffers from the effects of damp – spalling and efflorescence – and related defects such as cracking and crumbling. Repairs can usually be made in much the same way as for brickwork and render, although there are some special considerations you should be aware of. If the damage is widespread, however, it is quite straightforward to resurface the concrete prior to decorating.

Sealing concrete

New concrete has a high alkali content and efflorescence can develop on the surface as it dries out. Do not use any finish other than a water-thinned paint until the concrete is completely dry. Treat efflorescence on concrete in the same manner as for brickwork.

A porous concrete wall should be waterproofed with a clear sealant on the exterior. Some reinforced masonry paints will cover bitumen satisfactorily, but it will bleed through most paints unless you prime it with a PVA bonding agent diluted 50 per cent with water. Alternatively, use an aluminium spirit-based sealer.

Cleaning dirty concrete

Clean dirty concrete as you would brickwork, but where a concrete drive or garage floor is stained with patches of oil or grease, apply a proprietary oil-and-grease remover. Soak up fresh spillages immediately with dry sand or sawdust to prevent them becoming permanent stains.

Binding dusty concrete

Concrete is trowelled when it is laid in order to give a flat finish. If this is overdone, cement is brought to the surface and when the concrete dries out this thin layer begins to break up within a short time, producing a loose, dusty surface. It is not worth your while applying any decorative finish to concrete in this condition.

Treat a concrete wall with stabilizing primer, but paint a dusty floor with a concrete-floor sealer.

Repairing cracks and holes

Rake out and brush away loose debris from cracks and holes in concrete. If the crack is less than 6mm (¼in) wide, open it up a little with a cold chisel so that it will accept a filling. Undercut the edges to form a lip so the filler will grip.

To fill a hole in concrete, add a fine aggregate such as gravel to the sand-and-cement mix. Make sure the fresh concrete sticks in shallow depressions by priming the damaged surface with 3 parts bonding agent : 1 part water. When the primed surface is tacky, trowel in the concrete and smooth it.

Treating spalled concrete

When concrete breaks up, or spalls, due to the action of frost, the process is accelerated as steel reinforcement is exposed and begins to corrode. Fill the concrete as described above, but paint the metalwork first with a rust-inhibitive primer. If spalling recurs, particularly in exposed conditions, protect the wall with a bitumen base coat and a compatible reinforced masonry paint.

Spalling concrete
Rusting metalwork causes concrete to spall.

REPAIRING A CONCRETE FLOOR

An uneven or pitted concrete floor must be made flat and level before you apply any form of floorcovering. You can do this fairly easily yourself, using a proprietary self-levelling screed, but first of all you must ensure the surface is free from dampness.

Testing for dampness
Do not lay any sheet floorcoverings or tiles (or apply a levelling screed) to a floor that is damp. If the floor is new it must incorporate a damp-proof membrane; however, a new floor should be left to dry out for six months before an impermeable covering is added.

If you suspect an existing floor is damp, make a simple test by laying a small piece of polyethylene on the concrete and sealing it all round with self-adhesive parcel tape. After one or two days, inspect it for any traces of moisture on the underside.

For a more accurate assessment, hire a moisture meter. This device will read the moisture content of a suspect surface when you hold its contact pins against it. If the moisture reading does not exceed 6 per cent, you can proceed with covering or levelling the floor.

If either test indicates treatment is necessary, paint the floor with a bitumen-based waterproofer. Prime the surface first with a slightly diluted coat, then brush on two full-strength coats, allowing each to dry between applications. If necessary you can then lay a self-levelling screed over the waterproofer.

Applying a self-levelling compound
Fill holes and cracks deeper than about 3mm (⅛in) by first raking them out and undercutting the edges (1), then filling them with mortar mix.

Self-levelling compound is supplied as a powder which you mix with water. Make sure the floor is clean and free from damp, then pour some of the compound in the corner that is furthest away from the door. Spread the compound with a trowel (2) until it is about 3mm (⅛in) thick, then leave it to seek its own level. Continue across the floor, joining the area of compound until the entire surface is covered. You can walk on the floor after about one hour without damaging it, but leave the compound to harden for a few days before laying permanent floorcovering.

● **Cement-based exterior filler**
As an alternative for patching holes and rebuilding broken corners in concrete, use a proprietary cement-based exterior filler. When mixed with water, the filler remains workable for 20 minutes. Just before it sets hard, smooth or scrape the filler level.

1 Rake out cracks

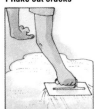

2 Apply compound
Spread self-levelling compound with a trowel.

● **Damp-proof membrane**
A continuous DPM of
polyethylene can be
inserted beneath the
concrete slab (see
right) or it can be
placed between the
slab and screed, in
which case the screed
should be at least
65mm (2½in) thick.
Alternatively, apply a
thick mastic DPC on
top of the concrete;
this can be in the form
of a bituminous
flooring adhesive.

REPAIRING A CONCRETE FLOOR

Concrete floors sometimes shrink and crack. Usually it is only the screed that has cracked and it can be repaired easily, but a cracked floor that is also uneven may be a sign of settlement in the sub-base and you should have it checked by a surveyor or by a Building Control Officer, who will advise you on what steps to take.

Filling a crack
Clean all dirt and loose material out of the crack and, if necessary, open up narrow parts with a cold chisel to allow better penetration of the filler.

Prime the crack with a solution of 1 part bonding agent : 5 parts water and let it dry. Make a filler of 3 parts sand : 1 part cement mixed with equal parts of bonding agent and water, or use a ready-mixed quick-setting cement. Apply the filler with a trowel, pressing it thoroughly into the crack.

Laying pipes in a concrete floor
House conversions or installations like central heating sometimes call for pipework to be run across a room. If the floor is solid that means either running it round the walls or setting it into the concrete. Although the latter method was common practice for plumbing, water bylaws now stipulate that pipes must not be embedded in a solid-concrete wall or floor. However, it is possible to conceal pipes in internal partition walls provided the water can be turned off in the event of a leak.

Another way to comply with these requirements is to run pipes inside moulded-plastic ducting laid in the solid floor. A plywood or a chipboard cover panel is screwed to the lipping of the duct to finish flush with the floor after the pipework has been fitted and tested. Any decorative floorcovering should ideally be loose-fitted or detailed to provide easy access.

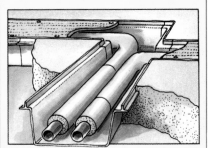

Plastic ducting for floor-run pipes

A suspended timber floor which has been seriously damaged by rot or insect infestation can be replaced with a solid concrete floor, provided the space below it needs no more than 600mm (2ft) of infill material. If it requires more, a concrete floor would be liable to damage through settlement of the infill, so a new suspended floor would have to be fitted.

Before taking any action consult your local Building Control Officer, because the converting of one floor can affect the ventilation of another and insulation may also be required.

If the work involves the electrical supply main or the supply pipes for gas or water you should check with the appropriate authority.

Wiring and heating pipes should be rerun before the infill is laid.

Preparing the ground

Strip out and burn all the old infected timbers and remove the door of the room. Treat the ground and all the surrounding masonry thoroughly with a strong fungicide. Fill in any recesses in the walls left by the timbers with bricks and mortar.

Mark the walls with a levelled chalk line to indicate the finished floor level, making allowance for the floorcovering if you intend to use a thick material such as quarry tiles or wood blocks. About 50mm (2in) below this line, mark another one, the space between them representing the thickness of the screed. Then mark a third chalk line a further 100mm (4in) down, indicating the thickness of the slab, followed by an allowance for 50mm (2in) thick polystyrene insulation board.

The infill

Lay the infill material to the required depth in layers of no more than 225mm (9in) at a time, compacting each layer thoroughly and breaking larger pieces with a sledgehammer **(1)**. You can use brick and tile rubble or, better still, gravel rejects (coarse stones from quarry waste). If you are using second-hand rubble, discard any fragments of plaster (which can react unfavourably with cement) and pieces of wood.

Bring the surface up to within 25mm (1in) of the chalk line for the insulation and 'blind' the surface with a layer of sand, tamped or rolled flat.

Spread a polyethylene damp-proof membrane of 1000-gauge (0.010in) minimum thickness over the surface of the sand, turning its edges up all round and lapping it up the walls to form a tray. Make neat folds at the corners and hold them temporarily in place with paper clips. If the floor needs more than one sheet of polyethylene to cover it, the sheets must overlap by at least 200mm (8in) and the joints should be sealed with a special waterproof tape available from builders' merchants.

1 Preparing the ground
Mark the walls with chalk lines for the finished floor level, thickness of the screed, thickness of the concrete slab and insulation. Fill the floor area with hardcore to within 25mm (1in) of the first line, compacting it well with a sledgehammer. Cover the hardcore with sand up to the line and lay a damp-proof membrane over it.

SEE ALSO
Details for:
Repairing concrete   29

Laying the concrete

Lay closely butted insulation board on the DPM and tape the joints. Provide strips of insulant around the edges up to screed level. Mix a medium-strength concrete of 1 part cement : 2½ parts sand : 4 parts aggregate. Do not add too much water; the mix should be a relatively stiff one.

Lay the concrete progressively in bands about 600mm (2ft) wide. The direction of the bands will depend on the door because you will have to work in such a way as to finish at the doorway. Tamp the concrete with a length of 100 x 50mm (4 x 2in) timber to compact it and finish level with the chalked line (2). As you go along, check the overall surface with a spirit level and straightedge and fill in any hollows, though slight unevenness will be taken up by the screed. Leave it to cure for at least three days under a sheet of polyethylene to prevent shrinkage caused by rapid drying.

Laying the screed

Mix the screed from 3 parts sharp sand : 1 part Portland cement. Dampen the floor and prime with a cement grout mixed to a creamy consistency with water and bonding agent in equal parts. Working from one wall, apply a 600mm (2ft) band of grout with a stiff brush.

Apply a bedding of mortar at each end of the grouted area to take 38 x 38mm (1½ x 1½in) 'screed battens'. True them with a spirit level and straightedge so that they are flush with the surface-level line on the walls.

Lay mortar between the battens and tamp it down well (3). Level the mortar with a straightedge laid across the battens, then smooth it with a wooden float. Lift out the battens carefully, fill the hollows with mortar and level with the float.

Repeat the procedure, working your way across the floor in bands 600mm (2ft) wide. Cover the finished floor with a sheet of polyethylene and leave it to cure for about a week. As soon as the floor is hard enough to walk on, trim the damp-proof membrane to within 25mm (1in) of the floor and fit the skirtings to cover its edges (4).

The floor will not be fully dry for about six months. Allow a month for every 25mm (1in) of thickness and in the meantime do not lay an impermeable floorcovering.

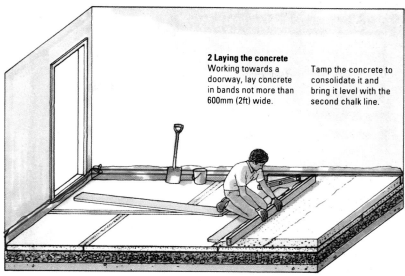

2 Laying the concrete
Working towards a doorway, lay concrete in bands not more than 600mm (2ft) wide.

Tamp the concrete to consolidate it and bring it level with the second chalk line.

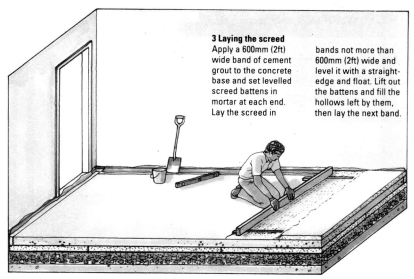

3 Laying the screed
Apply a 600mm (2ft) wide band of cement grout to the concrete base and set levelled screed battens in mortar at each end. Lay the screed in

bands not more than 600mm (2ft) wide and level it with a straight-edge and float. Lift out the battens and fill the hollows left by them, then lay the next band.

- **Insulating the floor**
  The degree of thermal insulation required varies according to the area of the floor and its construction. Check with your local BCO or ask an architect to calculate whether insulation is required.

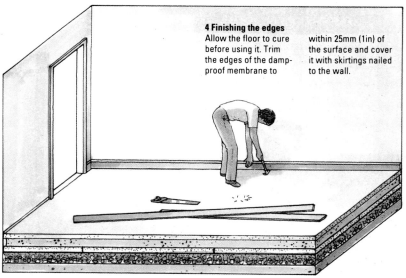

4 Finishing the edges
Allow the floor to cure before using it. Trim the edges of the damp-proof membrane to

within 25mm (1in) of the surface and cover it with skirtings nailed to the wall.

CLEANING BRICK AND STONE

Before you decorate the outside of your house, check the condition of the brick and stonework and carry out any necessary repairs. Unless you live in an area of the country where there is a tradition of painting brick and stonework, you will probably want to restore painted masonry to its original condition. Although most paint strippers cannot cope with deeply textured surfaces, there are thick-paste paint removers that will peel away layers of old paint from masonry walls.

Stained brickwork

Organic growth

Efflorescence

Treating new masonry

New brickwork or stonework should be left for about three months until it is completely dry before any further treatment is considered. White powdery deposits called efflorescence may come to the surface over this period, but you can simply brush them off with a stiff-bristle brush or a piece of dry sacking. Masonry is weatherproof and therefore requires no further treatment, except that in some areas of the country you may wish to apply paint.

Cleaning organic growth from masonry

There are innumerable species of mould growth and lichens which first appear as tiny coloured specks or patches on masonry. They gradually merge until the surface is covered with colours that range from bright orange to yellow, green, grey and black.

Moulds and lichen will only flourish in damp conditions, so try to cure the source of the problem before treating the growth. If one side of the house always faces away from the sun, for example, it will have little chance to dry out. Relieve the situation by cutting back any overhanging trees or shrubs to increase ventilation to the wall.

Make sure the damp-proof course (DPC) is working adequately and is not being bridged by piled earth or debris.

Cracked or corroded rainwater pipes leaking on to the wall are another common cause of organic growth. Feel behind the pipe with your fingers or slip a hand mirror behind it in order to locate the leak.

Removing the growth

Brush the wall vigorously with a stiff-bristle brush. This can be an unpleasant, dusty job, so wear a facemask. Brush away from you to avoid debris being flicked into your eyes.

Microscopic spores will remain even after brushing. Kill these with a solution of bleach or, if the wall suffers from persistent fungal growth, use a proprietary fungicide, available from most DIY stores.

Using a bleach solution

Mix 1 part household bleach with 4 parts water and paint the solution on to the wall with an old paintbrush. Wash the surface with clean water, using a scrubbing brush, 48 hours later. Brush on a second application of bleach solution if the original fungal growth was severe.

Using a fungicidal solution

Dilute the fungicide with water according to the manufacturer's instructions and apply it liberally to the wall with an old paintbrush. Leave it for 24 hours, then rinse the wall with clean water. In extreme cases, give the wall two washes of fungicide, allowing 24 hours between applications and a further 24 hours before washing it down with water.

Removing efflorescence from masonry

Soluble salts within building materials such as cement, brick, stone and plaster gradually migrate to the surface along with the water as a wall dries out. The result is a white crystalline deposit called efflorescence.

The same condition can occur on old masonry if it is subjected to more than average moisture. Efflorescence itself is not harmful, but the source of the damp causing it must be identified and cured before decoration proceeds.

Regularly brush the deposit from the wall with a dry stiff-bristle brush or coarse sacking until the crystals cease to form. Do not attempt to wash off the crystals – they will merely dissolve in the water and soak back into the wall. Above all, do not attempt to decorate a wall which is still efflorescing, because this is a sign that it is still damp.

When the wall is completely dry, paint the surface with an alkali-resistant primer to neutralize the effect of the crystals before you apply an oil paint. Masonry paints and clear sealants that let the wall breathe are not affected by the alkali content of the masonry, so can be used without applying a primer.

CLEANING OLD MASONRY

Improve the appearance of unpainted masonry by washing it with clean water. Starting at the top of the wall, play a hose gently on to the masonry while you scrub it with a stiff-bristle brush (1). Scrub heavy deposits with half a cup of ammonia added to a bucketful of water, then rinse again.

Removing unsightly stains

Soften tar, grease and oil stains by applying a poultice made from fuller's earth or sawdust soaked in paraffin or a proprietary grease solvent. Follow manufacturers' instructions and wear protective gloves.

Dampen the stain with solvent, then spread on a 12mm (½in) thick layer of poultice and leave it to dry out and absorb the stain. Scrape off the dry poultice with a wooden or plastic spatula and scrub the wall with water.

Stripping spilled paint

Remove a patch of spilled paint from brickwork with a proprietary paint stripper. Follow manufacturers' recommendations with regard to protective clothing. Stipple the stripper on to the rough texture (2). After about 10 minutes, remove the softened paint with a scraper and scrub the residue out of the deeper crevices with a stiff-bristle brush and water. Finally, rinse the wall with clean water.

1 Remove dirt and dust by washing

2 Stipple paint stripper on to spilled paint

A combination of frost action and erosion tends to break down the mortar pointing between bricks and stonework. The mortar eventually falls out, exposing the open joints to wind and rain that eventually drive dampness through the wall to the inside. Replacing pointing is straightforward, but time-consuming. Tackle a small, manageable area at a time, using a ready-mixed mortar or your own mix.

Applying the mortar

Rake out the old pointing with a thin wooden lath to a depth of about 12mm (½in). Use a cold chisel or a special plugging chisel and a club hammer to dislodge sections that are firmly embedded, then brush out the joints with a stiff-bristle brush.

Spray the wall with water to make sure the bricks or stones will not absorb too much moisture from the fresh mortar. Mix up some mortar in a bucket and transfer it to a hawk. If you are mixing your own mortar, use the proportions 1 part cement : 1 part lime : 6 parts builders' sand.

Pick up a little sausage of mortar on the back of a small pointing trowel and push it firmly into the upright joints. This can be difficult to do without the mortar dropping off, so hold the hawk under each joint to catch it. Try not to smear the face of the bricks with mortar as it will stain. Repeat the process for the horizontal joints. The actual shape of the pointing is not vital at this stage.

Once the mortar is firm enough to retain a thumbprint it is ready for shaping. Match the style of pointing used on the rest of the house (see below). When the pointing has almost hardened, brush the wall to remove traces of surplus mortar.

Shaping the mortar joints

The joints shown here are commonly used for brickwork. Rubbed joints are best for most stonework. Leave the pointing of dressed-stone ashlar blocks to an expert.

Flush joints

The easiest profile to produce, a flush joint is stippled with a stiff-bristle brush to expose the sand aggregate.

Rubbed (concave) joints

Bricklayers make a rubbed or rounded joint with a tool shaped like a sled runner with a handle: the semi-circular blade is run along the joints. Improvise by bending a length of metal tube or rod; use the curved section only or you will gouge the mortar. Once the mortar is shaped, stipple it so that it matches the weathered pointing.

Raked joints

A raked joint is used to emphasize the type of bonding pattern of a brick wall. It is not suitable for soft bricks or for a wall that takes a lot of weathering. Scrape out a little of the mortar, then tidy up the joints by running a 9mm (⅜in) lath along them.

Weatherstruck joints

The sloping profile is intended to shed rainwater from the wall. Shape the mortar with the edge of a pointing trowel. Start with the vertical joints and slope them in either direction, but be consistent. During the process, mortar will tend to spill from the bottom of a joint as surplus is cut off. Bricklayers use a tool called a 'Frenchman' to neaten the work: it has a narrow blade with the tip bent at right angles. Make your own tool by bending a thin metal strip and binding insulating tape round the other end to form a handle, or bend the tip of an old kitchen knife after heating it in the flame of a blowtorch or cooker burner.

You will find it easiest to use a wooden batten to guide the blade of the Frenchman along the joints, but nail scraps of plywood at each end of the batten to hold it off the wall. Align the batten with the bottom of the horizontal joints, then draw the tool along it, cutting off the excess mortar.

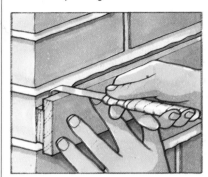

Use a Frenchman to trim weatherstruck joints

SEE ALSO
Details for:
Penetrating damp   7-9

● **Mortar dyes**
Liquid or powder additives are available for changing the colour of mortar to match existing pointing. Colour matching is difficult and smears can stain the bricks permanently.

Flush joint

Rubbed joint

Raked joint

Weatherstruck joint

33

REPAIRING
MASONRY

Cracked masonry may simply be the result of cement-rich mortar being unable to absorb slight movements within the building. However, it could also be a sign of a more serious problem – subsiding foundations, for example. Do not just ignore the symptoms but investigate immediately and put the necessary repairs in hand.

Filling cracked masonry

If a brick or stone wall has substantial cracks, consult a builder or your local Building Control Officer to ascertain the cause. If a crack proves to be stable you can carry out repairs yourself.

Cracked mortar can be removed and repointed in the normal way, but a crack that splits one or more bricks cannot be repaired neatly and the damaged masonry should be replaced in the same manner as spalled brickwork (see right).

Cracks across a painted wall can be filled with mortar that has been mixed with a little PVA bonding agent to help it to stick. Before you effect the repair, wet the damaged masonry with a hose to encourage the mortar to flow deeply into the crack.

Crack may follow pointing only

Cracked bricks could signify serious faults

Priming brickwork for painting

Brickwork will only need to be primed if it is showing signs of efflorescence or spalling. An alkali-resistant primer will guard against the former and a stabilizing solution will both bind crumbling masonry and help to seal it.

If you are planning to paint the wall for the first time with masonry paint, you may find that the first coat is difficult to apply due to the suction of the dry, porous brick. Thin the first coat slightly with water or solvent.

Waterproofing masonry

1 Replacing a spalled brick
Having mortared the top and one end, slip the new brick into the hole you have cut.

Colourless water-repellent fluids are intended to make masonry impervious to water without colouring it or stopping it from breathing, which is important in order to allow moisture within the walls to dry out.

Prepare the surface thoroughly before applying the fluid: make good any cracks in bricks or pointing and remove organic growth, then allow the wall to dry out thoroughly.

Apply the fluid generously with a large paintbrush and stipple it into the joints. Apply a second coat as soon as the first has been absorbed to ensure that there are no bare patches where water could seep in. So that you can be sure you are covering the wall properly,

use a sealant containing a fugitive dye which will disappear gradually after a specified period.

Carefully paint up to surrounding woodwork; if you accidentally splash sealant on to it, wash it down immediately with a cloth dampened with white spirit.

The fumes from the fluid can be dangerous if inhaled, so be sure to wear a proper respirator as recommended by the manufacturer. If you need to treat a whole house, it might be worth hiring a company which will undertake spraying on the sealant. Make sure their workmen rig up plastic-sheet screens to prevent overspray drifting across to your neighbour's property.

REPAIRING SPALLED MASONRY

Moisture that has penetrated soft masonry will expand in icy weather, flaking off the outer face of brickwork and stonework. The process, known as spalling, not only looks unattractive but also allows water to seep into the surface. Repairs to spalled bricks or stones can be made, though treatment depends on the severity of the problem.

If spalling is localized, cut out the bricks or stones and replace them with matching ones. The sequence below describes doing this with brickwork, but the process is similar for a stone wall.

Spalled bricks caused by frost damage

Where spalling is extensive, the only practical solution is to accept its less-than-perfect appearance, repoint the masonry and apply a clear water repellent that will protect the wall from any further damage while at the same time allowing it to breathe.

Replacing a spalled brick
Use a cold chisel and club hammer to rake out the pointing surrounding the brick, then chop out the brick itself. If the brick is difficult to prise out, drill numerous holes in it with a large-diameter masonry bit, then slice up the brick with a cold chisel and hammer: it should crumble, enabling you to remove the pieces easily.

To fit the replacement brick, first dampen the opening and spread mortar on the base and one side. Butter the dampened replacement brick with mortar on the top and one end and slot it into the hole (**1**). Shape the pointing to match the surrounding brickwork.

If you can't find a replacement brick of a suitable colour, remove the spalled brick carefully, turn it round to reveal its undamaged face and reinsert it.

REPAIRING RENDER

Brickwork may be clad with a smooth or roughcast cement-based render, both for improved weatherproofing and to give a decorative finish. Often the render is susceptible to the effects of damp and frost, which can cause cracking, bulging and staining. Before you redecorate a rendered wall make good any damage and clean off surface dirt, mould growth and flaky material in order to achieve a long-lasting finish.

Cracked render allows moisture to penetrate

Blown pebbledash parts from the masonry

Leaky guttering causes rust stains

Repairing defective render

Before you repair cracked render, correct any structural faults that may have contributed to it. Apply a stabilizing solution if the wall is dusty.

Ignore fine hairline cracks if you intend to paint the wall with a reinforced masonry paint. Rake out larger cracks with a cold chisel, dampen with water and fill them flush to the surface with an exterior filler. Fill any major cracks with a render made of 1 part cement : 2 parts lime : 9 parts builder's sand, plus a little PVA bonding agent to help it adhere to the masonry.

Bulges in render normally indicate that the cladding has parted from the masonry. Tap gently with a wooden mallet to find the extent of these hollow areas and hack off the material to sound edges. Undercut the perimeter of each hole except for the bottom edge, which should be left square.

Brush out debris, then apply a coat of PVA bonding agent. When it becomes tacky, trowel on a 12mm (½in) thick layer of 1 : 1 : 6 render and leave it to set firm. Scratch it to form a key and, the next day, fill flush with a slightly weaker 1 : 1 : 9 mix. Smooth the surface with a wooden float, using circular strokes.

Reinforcing a crack in render

To prevent a crack in render opening up again, reinforce the repair with a glass-fibre membrane embedded in a bitumen base coat. Rake out the crack to remove any loose material, then wet it. Fill just proud of the surface with a mortar mix of 1 part cement : 4 parts builders' sand. When this has stiffened scrape it flush with the render.

When the mortar has hardened, brush on a generous coat of bitumen base coat, making sure it extends at least 100mm (4in) on both sides of the crack. Embed strips of fibre-glass scrim (sold with the base coat) into the bitumen, using a stippling and brushing action (1). While it is still wet, feather the edges of the bitumen with a foam roller (2), bedding the scrim into it. After 24 hours the bitumen will be hard, black and shiny. Apply a second coat, feather with a roller and, when it has dried, apply two full coats of a compatible reinforced masonry paint.

1 Embed the scrim

2 Feather with roller

Patching pebbledash

Pebbledash comprises small stones stuck to a thin coat of render over a thicker base coat. If damp gets behind pebbledashing, one or both layers may separate. Hack off any loose render to a sound base and seal it with stabilizer. If necessary, repair the scratchcoat of render. Simulate the texture of pebbledash with a thick paste made from PVA bonding agent. Mix 1 part cement-paint powder with 3 parts clean sharp (plastering) sand. Stir in 1 measure of bonding agent diluted with 3 parts water to form a thick, creamy paste. Load a banister brush and scrub the paste on to the bare surface.

Apply a second generous coat of paste, stippling it to form a coarse texture. Leave for about 15 minutes to firm up then, with a loaded brush, stipple it to match the texture of the pebbles. Let the paste harden fully before painting.

To leave the pebbledash unpainted, make a patch using replacement pebbles. The result may not be a perfect match, but could save you having to paint the entire wall. Cut back the blown area and apply a scratchcoat followed by a buttercoat. While this is still wet, fling pebbles on to the surface from a dustpan; they should stick to the soft render, but you may have to repeat the process until the coverage is even.

Stipple the texture

Removing rust stains

Faulty plumbing will often leave rusty streaks on a rendered wall. Before decorating, prime the stains with an aluminium spirit-based sealer or they will bleed through. Rust marks may also appear on a pebbledashed wall, well away from any metalwork: these are caused by iron pyrites in the aggregate. Chip out the pyrites with a cold chisel, then seal the stain.

EXTERIOR
RENDERING

Rendering is the application of a relatively thin layer of cement or cement-and-lime mortar to the surfaces of exterior walls to provide a decorative and weather-resistant finish.

Any such treatment of exterior walls should be carefully considered beforehand, because the finished outer surface should always harmonize with the character of a building and not look ill at ease with those of its neighbours. This is particularly important in the case of terraced housing, where the fronts of the houses form an unbroken run of wall.

Planning ahead
Except for listed buildings or houses in a conservation area, there are no regulations controlling the change of colour or texture of exterior walls. Consequently, houses are often made conspicuous by their individualistic decorative treatment.

Re-rendering a wall is always acceptable as it is merely a case of renewing what is already there. Rendering old brickwork might improve its weather-resistance, but at the considerable cost of destroying the appearance of the building. Here it would be better to rake out the mortar joints, repoint them and, if necessary, treat the brickwork with a clear sealant.

Rendering techniques
The technique used in rendering is virtually the same as in plastering, for which cement and lime are also sometimes used. It generally involves using the same tools, though a wooden float is better than a plasterer's trowel for finishing cement rendering. The wood leaves a finely textured surface that looks better than the very smooth one produced with a metal trowel.

Rendering the walls of a house is really a job for a professional, as it involves covering a large area evenly, and also requires the ability to colour-match the batches of mortar, which is critical if the finished job is not to look patchy.

While a non-professional can undertake repairs to rendering, it is still difficult to match the colour of the new work to the old. You might consider painting the rendered wall.

For new work, divide the wall into manageable panels with screed battens as for plastering. However, colour-matching the mortar will still be a problem, and 'losing' the joints can be difficult. It might pay to concentrate on getting the rendering flat and then disguising any patchiness with paint, but once again you should consider the character of the house.

Before attempting to render a large wall, practise if possible on a smaller project, such as a garden wall.

BINDERS FOR EXTERIOR RENDERING

Mortar

Mortar is a mixture of sand, cement and clean water. The sand gives the mix bulk and the cement binds the particles. A cement mix will bond to any ordinary masonry material and to metals. Mortars of various strengths are produced by adjusting the proportions of sand and cement, or by adding lime.

A mortar should not be stronger than the materials on to which it is being applied. A cement-and-sand mix can be applied to a wall of dense hard bricks, for example, but a weaker mix of cement, sand and lime is more appropriate for soft bricks or blocks.

Cement sets by a chemical reaction with water known as hydration, and begins as soon as the water is added. Cement does not need to dry out in order to set, and the more slowly it dries out the stronger it will be.

Normally an average mix will stay workable for at least two hours. It will continue to gain strength for a few days after its initial set, reaching full strength in about a month.

Hot weather will reduce the workable time and can affect the set of the mortar by making it dry out too fast. In these conditions the work should be kept damp by being lightly sprayed with water or by being covered with polyethylene sheeting to retain the moisture and slow the drying time.

Cement

Cement made from limestone or chalk and clay is generally called Portland cement. Various types are made by adding other materials or by modifying the production methods.

Ordinary Portland cement (OPC)
This common light grey cement is mixed with aggregates for concrete and mortars. It is available in 50kg (110lb) bags and in smaller amounts.

Sulphate-resistant Portland cement
This is used in areas where soluble sulphates cause degrade problems.

White Portland cement
Similar to ordinary Portland but is more expensive. It makes light-coloured mixes for bricklaying, concrete and rendering. Pigment powders are available for colouring mortar mixes. The materials must be carefully proportioned for the batches to match.

Quick-setting cement
This cement is mixed with water and sets hard in 30 minutes. It is non-shrinking and waterproof and is useful for small repair jobs.

Masonry cement
This grey cement is specially made for rendering and bricklaying.

Lime

Lime is made from limestone or chalk. When it leaves the kiln it is called quicklime, and may be non-hydraulic or hydraulic. Non-hydraulic lime, in general use, sets by combining with carbon dioxide from the air as the water mixed with it dries out. Hydraulic lime has similar properties to cement; it sets when water is added and so can be used under water. When quicklime – the non-hydraulic type especially– is 'slaked' (mixed with water) it expands and gives off heat.

Lime must be properly slaked before use, and at one time a batch would have been soaked in a tub for weeks before it was used. This soaked lime was called lime putty.

Preslaked non-hydraulic lime powder, or hydrated lime, is sold by builders' merchants. It can be used at once, but is often soaked for 24 hours before use to make lime putty. The lime is mixed with water to a creamy consistency or with sand and water and left to stand as lime mortar called 'coarse stuff'. This can be kept for some days without setting if it is heaped up and covered with polyethylene sheet to prevent the water evaporating.

The less active hydraulic lime is dry-mixed with the sand like cement powder and the slaking process takes place when water is added.

Mortars are mixed with the finest aggregate, sand. Sand is graded by the size and shape of its particles; a well-graded sand will have particles of different sizes rather than ones that are uniformly large or small.

Types of sand

Sharp sand is used with coarse aggregates for making concrete and floor screeds. Plasterers' sharp sand is of a finer grade, and is used for rendering. Builders' sand, also known as bricklayers' or soft sand, has smoother particles and is used for masonry work. Use well-washed sand, as impurities can weaken a mortar and affect the set. A good sand should not stain your hand if you squeeze it.

Most aggregates may be bought from builders' merchants by the cubic metre; some suppliers sell it in small packs.

Stone chippings

Specially prepared crushed stone in various colours is used for pebble-dash rendering. Order enough for the whole job in hand (your supplier should be able to advise you) as additional stones bought later may not match the colour of the original batch. If you do run short, stop work at a corner rather than partway across a wall. The extra stones can be mixed with the remaining ones and a subtle change of colour is less likely to show on the adjacent wall.

Dry-mixed mortar

Prepacked sand-and-cement mortar mixes are sold by builders' merchants and DIY shops in large and small packs. They are ready-proportioned for different kinds of application and require only water to be added. As sand and cement 'settle out', a whole bag should be used and mixed well before adding the water.

Premixed and self-coloured one-coat renders are also available as an alternative to the traditional two-coat variety. The available colours are white, ivory, cream, stone, grey and pink.

STORING SAND AND CEMENT

Storing the materials should not normally be necessary because it is best to buy them as required and use them up by the end of the job.

However, if you are held up for a time after taking delivery, store powder or premixed materials as recommended for plaster. Store sand in a neat heap on a board or a plastic sheet and protect it from windblown dirt and rain with plastic sheeting.

Storing sand
Dirty sand can affect the set of the cement. Keep it covered with plastic sheeting.

ADDITIVES

Proprietary additives which modify the properties of mortar are added to the mix in precise proportions according to manufacturers' instructions. Their functions vary. Waterproofers, which make mortar impervious by sealing its pores, may be used when rendering on exposed walls. Plasticizers (additives which make a mortar easier to work) can be used instead of lime.

SEE ALSO
Details for:
Mixing mortar   38

MORTAR MIXES FOR TWO-COAT RENDERING

The mix for a mortar will depend upon the strength of the material that is being rendered as well as the degree of exposure. The mix for the undercoat should not be stronger than the background, and the top coat should be no stronger than the undercoat.

Though these considerations are hardly ever critical for the majority of DIY work, when a situation does dictate that a precise mix is required, the proportions of the materials must be measured quite accurately.

Measure the material by volume, using a bucket. Loose cement and damp sand tend to 'bulk up' when loaded. Encourage cement powder to settle by tapping the bucket, and then top it up. Damp sand will not settle, so increase the measure by 25 per cent. Dry sand and saturated sand will settle to a normal measure.

TYPE OF BACKGROUND	TYPE OF MIX	PARTS BY VOLUME			
		AVERAGE EXPOSURE		SEVERE EXPOSURE	
		UNDERCOAT	TOP COAT	UNDERCOAT	TOP COAT
Low-suction backgrounds: Hard, dense clay bricks Dense concrete blocks Stone masonry Normal ballast concrete	Cement : lime : sand	1 : ½ : 4½	1 : 1 : 6*	1 : ½ : 4½	1 : 1 : 6
	Cement : sand & plasticizer	1 : 4	1 : 6*	1 : 4	1 : 6
	Masonry cement : sand	1 : 3½	1 : 5*	1 : 3½	1 : 5
Normal-absorption backgrounds: Most average-strength bricks Clay blocks Normal concrete blocks Aerated concrete blocks	Cement : lime : sand	1 : 1 : 6	1 : 2 : 9*	1 : 1 : 6	1 : 1 : 6
	Cement : sand & plasticizer	1 : 6	1 : 8*	1 : 6	1 : 6
	Masonry cement : sand	1 : 5	1 : 6½*	1 : 5	1 : 5

*Suitable mixes for interior plastering undercoats

37

MIXING MORTAR FOR RENDERING

1 Shovel dry mix from one heap to another

2 Form a well in the heap and pour in water

3 Shovel dry material into water

4 Sprinkle mix with water if too dry

5 Test consistency of mix with shovel

Mix only as much mortar as you can use in an hour and, if the weather is very hot and dry, shorten this to half an hour. Keep all your mixing tools and equipment thoroughly washed so that no mortar sets on them.

Measure the required level bucketfuls of sand on to the mortar board or, for larger quantities, on to a smooth, level base such as a concrete drive.

Using a second dry bucket and shovel, kept exclusively for cement powder, measure out the cement, tapping the bucket to settle the loose powder and topping it up as needed. Tip the cement over the heaped-up sand and mix the sand and cement together by shovelling them from one heap to another and back again (1). Continue to turn this dry mix (the sand will actually be damp) until the whole takes on a uniform grey colour.

Form a well in the centre of the heap and pour in some water (2) – but not too much at this stage.

Shovel the dry mix from the sides of the heap into the water until the water is absorbed (3). If you are left with dry material, add more water as you go until you achieve the right firm, plastic consistency in the mortar, turning it repeatedly to mix it thoroughly to an even colour. It is quite likely that you will misjudge the amount of water at first, so if after turning the mix is still relatively dry, sprinkle it with water (4). Bear in mind, though, that too much water will weaken the mix.

Draw the back of your shovel across the mortar with a sawing action to test its consistency (5). The ribs formed in the mixture should not slump back or crumble. That would indicate that the mortar is either too wet or too dry respectively. The back of the shovel should leave a smooth texture on the surface of the mortar.

Make a note of the amount of water used in proportion to the dry materials so that further mixes will be consistent.

For cement-lime-sand mixes, the lime powder can be added with the cement and dry-mixed as described above. Otherwise lime putty can be mixed with the sand before the cement is added, or the cement can be added to prepared 'coarse stuff'. When you have finished, hose down and sweep clean the work area, particularly if it is a driveway, as any remaining cement slurry will stain the surface.

MIXING MORTAR BY MACHINE

You can hire a small-capacity electric or petrol-driven cement mixer. Such a mixer can save you a great deal of time and effort, especially on big jobs, and is quite easy to use.

Set the machine as close as possible to the work area and place a board under the drum to catch any spilt materials. If it is an electric-powered machine, take all due precautions with the power supply and keep the cables well clear of the work.

Load the drum with half the measure of sand and add a similar proportion of cement, and lime if required. Dry-mix them by running the mixer, then add some water.

Load the remainder of the materials in the same sequence, adding a little water in between.

Run the mixer for a couple of minutes to mix the materials thoroughly, then stop the machine and test some of the mix for consistency.

Generally, it pays to make a rendering mix somewhat stiffer for blockwork than for brickwork. However, this depends to some extent on the absorbency of the background.

It is advisable to wash out the drum of the mixer after each mix and to scour it out with water and some coarse aggregate at the end of the working day. If you return the machine with dry or drying mortar in its drum you may be charged extra.

Cement mixer
Hire an electric or petrol-driven mixer when a large batch of mortar is required.

TEXTURED RENDERINGS

It is possible to use tools to texture rendering while it is still damp, but it is more usual to apply a coarse aggregate. This is a fairly skilled procedure (see below for details). Try to reproduce a texture when patch-repairing.

Roughcast rendering
For this rendering, mix aggregate no more than 9mm (⅜in) in size with the top-coat mortar. Add about half as much as the amount of sand used, with enough water for a sticky mix. Flick it on the wall to build up an even coat.

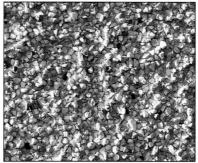

Pebble-dash rendering
Crushed-stone aggregate gives pebble dash its colour, and an even distribution of the chippings is necessary in order to avoid patchiness. A 9 to 12mm (⅜ to ½in) top coat is applied and the stones are thrown at it while it is soft, then pressed with a float to bed them in.

Tyrolean finish
A fine cement mix is sprayed from a hand-cranked 'Tyrolean machine' to build up a decorative texture over a dry undercoat rendering. Doors, windows, gutters and so on must be masked beforehand. Tyrolean machines may be hired.

Preparing the surface

Using a hammer and chisel, neatly chop away the old loose coating on areas of cracked or blistered rendering. Rake out the mortar joints in the exposed brickwork if necessary and brush the area down. Clean off any organic growth such as lichen or algae that may be present and apply a fungicide.

Work platform

Set up a safe work platform from which to do the rendering. You will need both hands free to use the tools, so it cannot be done from a ladder. Pairs of steps with a scaffold board between them can be used for working on ground-floor walls, but for upper walls you will need a scaffold tower.

Two-coat work

Set up 9mm (⅜in) vertical screed battens spaced no more than 900mm (3ft) apart, fixing them with masonry nails into the mortar joints of the brickwork. Check them for level and pack them out where necessary.

Apply undercoat rendering between two battens, using a firm pressure to make it bond well on to the dampened background. Build up the rendering to the thickness of the screed battens.

Level the mortar with a straightedge laid across the battens, working upward with a side-to-side movement, then scratch the surface of the mortar to provide a key for the top coat and leave it to set for a week.

You can fill in the panels between the battens in sequence or alternately. Allow the rendering to set before you remove the battens.

Apply top-coat rendering about 6mm (¼in) thick, either freehand or with the aid of screed battens as before. Use a straightedge for levelling the render, then finish it with a wooden float.

One-coat work

Set out 16mm (⅝in) vertical screed battens, spaced as for two-coat work. Apply the render to the brickwork in a single coat. Level the surface and allow the render to stiffen, then finish with a wooden float or texture the surface with a proprietary scraper, working it in a circular motion.

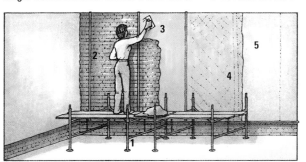

Applying two-coat rendering
1 Set up a safe work platform.
2 Divide the wall with vertical screeds.
3 Apply the undercoat between all screeds or on alternate panels.
4 Remove screeds and fill in gaps or panels.
5 Apply top coat over keyed undercoat.

Patch repairs

Use a metal plasterer's trowel to apply the rendering and finish the top coat with a wooden float.

Take a trowelful of mortar from your hawk and spread it on the wall with an upward stroke, applying firm pressure **(1)**. Level the surface of a one-coat render with a straightedge laid across the surfaces of the surrounding rendering and worked upward with a side-to-side motion. Finish with a float.

For two-coat work, build up the undercoat layer no more than two-thirds the thickness of the original rendering or 9mm (⅜in), whichever is the thinner.

Level the mortar with a straightedge that fits within the cut-out of the area being patched, then key the surface for the top coat **(2)**. Leave the undercoat to set and strengthen for a few days.

Before applying the top coat, dampen the undercoat rendering to even out the suction. Finally, level the top coat with a straightedge and finish with a float.

1 Use firm pressure

2 Key the surface

PAINTING
EXTERIOR
MASONRY

SEE ALSO
Details for:
Preparing masonry   32-35

Strain old paint
If you're using leftover paint, filter it through a piece of muslin or old tights stretched over the rim of a container.

Resealing the lid
Wipe the rim of the can clean before you replace the lid, then tap it down all round with a hammer over a softwood block.

SAFETY WHEN PAINTING

DECORATING WITH SOLVENT-BASED PAINT IS NOT DANGEROUS PROVIDED YOU TAKE SENSIBLE PRECAUTIONS.

● Ensure good ventilation indoors while applying a finish and when it is drying. Wear a respirator if you suffer from breathing disorders.
● Do not smoke while painting or in the vicinity of drying paint.
● Contain paint spillages outside with sand or earth and don't allow any paint to enter a drain.
● If you splash paint in your eyes, flush them with copious amounts of water with your lids held open; if symptoms persist, visit a doctor.
● Always wear barrier cream or gloves if you have sensitive skin. Use a proprietary skin cleaner to remove paint from your skin or wash it off with warm soapy water. Do not use paint thinners to clean your skin.
● Keep any finish and thinners out of reach of children. If a child swallows a substance, do not make any attempt to induce him or her to vomit – seek medical treatment instead.

PREPARING THE PAINT

Whether you're using newly purchased paint or leftovers from previous jobs, there are some basic rules to observe before you apply it.
● Wipe dust from the paint can, then prise off the lid with the side of a knife blade. Don't use a screwdriver: it only buckles the edge of the lid, preventing an airtight seal and making subsequent removal difficult.
● Gently stir liquid paints with a wooden stick to blend the pigment and medium. There's no need to stir thixotropic paints unless the medium has separated; if you have to stir it, leave it to gel again before using.
● If a skin has formed on paint, cut round the edge with a knife and lift out in one piece with a stick. It's a good idea to store the can on its lid, so that a skin cannot form on top of the paint.
● Whether the paint is old or new, transfer a small amount into a paint kettle or plastic bucket. Old paint should be filtered at the same time, tying a piece of muslin or old nylon tights across the rim of the kettle.

The outside walls of houses are painted for two major reasons: to give a clean, bright appearance and to protect the surface from the rigours of the climate. What you use as a finish and how you apply it depends on what the walls are made of, their condition and the degree of protection they need. Bricks are traditionally left bare, but may require a coat of paint if previous attempts to decorate have resulted in a poor finish. Rendered walls are often painted to brighten the naturally dull grey colour of the cement; pebbledashed surfaces may need a colourful coat to disguise previous conspicuous patches. On the other hand, you may just want to change the present colour of your walls for a fresh appearance.

Working to a plan

Before you embark upon painting the outside walls of your house, plan your time carefully. Depending on the amount of preparation that is required, even a small house will take a few weeks to complete.

It is not necessary to tackle the whole job at once – although it is preferable, as the weather may change to the detriment of your timetable. You can split the work into separate stages with days (even weeks) in between, provided you divide the walls into manageable sections. Use window frames and doorframes, bays, downpipes and corners of walls to form break lines that will disguise joins.

Start at the top of the house, working from right to left if you are right-handed and vice versa.

● Black dot denotes compatibility.
All surfaces must be clean, sound, dry and free from organic growth.

FINISHES FOR MASONRY

	Cement paint	Water-based masonry paint	Reinforced masonry paint	Solvent-based masonry paint	Textured coating	Floor paint
SUITABLE TO COVER						
Brick	●	●	●	●	●	●
Stone	●	●	●	●	●	●
Concrete	●	●	●	●	●	●
Cement rendering	●	●	●	●	●	●
Pebbledash	●		●		●	
Emulsion paint		●	●	●	●	●
Solvent-based paint		●	●	●	●	●
Cement paint	●	●	●	●	●	●
DRYING TIME: HOURS						
Touch dry	1–2	1–2	2–3	4–6	6	2–3
Recoatable	24	4–6	24	16	24–48	3–16
THINNERS: SOLVENTS						
Water	●	●	●		●	●
White spirit				●	●	●
NUMBER OF COATS						
Normal conditions	2	2	1–2	2	1	1–2
COVERAGE: DEPENDING ON WALL TEXTURE						
Sq metres per litre		4–10	3–6.5	6–16	2	5–10
Sq metres per kg	1–6				1–2	
METHOD OF APPLICATION						
Brush	●	●	●	●	●	●
Roller	●	●	●	●	●	●
Spray gun	●	●	●	●		

SUITABLE PAINTS FOR EXTERIOR MASONRY

There are various grades of paint suitable for decorating and protecting exterior masonry which take into account economy, standard of finish, durability and coverage. Use the chart opposite for quick reference.

Cement paint

Cement paint is supplied as a dry powder, to which water is added. It is based on white cement, but pigments are added to produce a range of colours. Cement paint is one of the cheaper paints suitable for exterior use. Spray new or porous surfaces with water before applying two coats.

Mixing cement paint

Shake or roll the container to loosen the powder, then add 2 volumes of powder to 1 of water in a clean bucket. Stir it to a smooth paste, then add a little more water until you achieve a full-bodied, creamy consistency. Mix up no more than you can use in one hour, or it will start to dry.

Adding an aggregate

When you are painting a dense wall or one treated with a stabilizing solution so that its porosity is substantially reduced, it is advisable to add clean sand to the mix to give it body. It also provides added protection for an exposed wall and helps to cover dark colours. If the sand changes the colour of the paint, add it to the first coat only. Use 1 part sand to 4 parts powder, stirring it in when the paint is still in its paste-like consistency.

Masonry paints

When buying weather-resistant exterior-masonry paints you have a choice between a smooth matt finish or a fine granular texture.

Water-based masonry paint

Most masonry paints are water-based, being in effect exterior-grade emulsions with additives that prevent mould growth. They are supplied ready for use, but in fact it pays to thin the first coat on porous walls with 20 per cent water. Follow up with one or two full-strength coats, depending on the colour of the paint.

Water-based masonry paints must be applied during fairly good weather. Damp or humid conditions and low temperatures may prevent the paint drying properly.

Solvent-based masonry paints

Some masonry paints are thinned with white spirit or with a special solvent, but unlike most oil paints they are moisture-vapour permeable so that the wall is able to breathe. It is often advisable to thin the first coat with 15 per cent white spirit, but check with the manufacturer's recommendations .

Solvent-based paints can be applied in practically any weather conditions, provided it is not actually raining.

Reinforced masonry paint

Masonry paint that has powdered mica or a similar fine aggregate added to it dries with a textured finish that is extremely weatherproof. Reinforced masonry paints are especially suitable in coastal districts and in industrial areas – where dark colours are also an advantage in that dirt will not show up so clearly as on a pale background. Although large cracks and holes must be filled prior to painting, reinforced masonry paint will cover hairline cracks and crazing.

Textured coating

A thick textured coating can be applied to exterior walls to form a thoroughly weatherproof, self-coloured coating which can also be overpainted to match other colours. The usual preparation is necessary and brickwork should be pointed flush. Large cracks should be filled, although a textured coating will cover fine cracks. The paste is brushed or rolled onto the wall, then left to harden, forming an even texture. However, if you prefer, you can produce a texture of your choice using a variety of simple tools. It is an easy process, but put in some practice on a small section first.

Concrete floor paints

Floor paints are specially prepared to withstand hard wear. They are especially suitable for concrete garage or workshop floors, but they are also used for stone paving, steps and other concrete structures. They can be used inside for playroom floors.

The floor must be clean, dry and free from oil or grease. If the concrete is freshly laid, allow it to mature for at least a month before painting. Prime powdery or porous floors with a proprietary concrete sealer.

The best way to paint a large area is to use a paintbrush around the edges, then fit an extension to a paint roller for the bulk of the floor.

Apply paint with a roller on an extension

SEE ALSO
Details for:
Preparing masonry.   32-35

Paint in manageable sections
You can't hope to paint an entire house in one session, so divide each elevation into manageable sections to disguise the joins. The horizontal moulding divides the wall neatly into two sections, and the raised door and window surrounds are convenient break lines.

41

TECHNIQUES
FOR PAINTING
MASONRY

SEE ALSO
Details for:
Preparing masonry   32–35

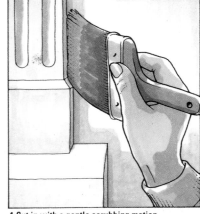

1 Cut in with a gentle scrubbing motion

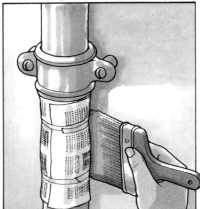

2 Protect downpipes with newspaper

3 Use a banister brush
Tackle deeply textured
wall surfaces with a
banister brush, using a
scrubbing action.

4 Use a roller
For speed in
application, use a
paint roller with a deep
pile for heavy textures
and a medium pile for
light textures and
smooth wall surfaces.

5 Spray onto the apex of external corners

6 Spray internal corners as separate surfaces

Using paintbrushes

Choose a 100 to 150mm (4 to 6in) wide
paintbrush for walls; larger ones are
heavy and tiring to use. A good-quality
brush with coarse bristles will last
longer on rough walls. For a good
coverage, apply the paint with vertical
strokes, criss-crossed with horizontal
ones. You will find it necessary to
stipple paint into textured surfaces.

Cutting in
Painting up to a feature such as a
doorframe or window frame is known
as cutting in. On a smooth surface, you
should be able to paint a reasonably
straight edge following the line of the
feature, but it's difficult to apply the
paint to a heavily textured wall with a
normal brush stroke. Don't just apply
more paint to overcome the problem;
instead, touch the tip of the brush only
to the wall, using a gentle scrubbing
action (1), then brush excess paint
away once the texture is filled.
   Wipe splashed paint from window
frames and doorframes with a cloth
dampened with the appropriate thinner.

Painting behind pipes
To protect rainwater downpipes, tape a
roll of newspaper around them. Stipple
behind the pipe with a brush, then slide
the paper tube down the pipe to mask
the next section (2).

Painting with a banister brush
Use a banister brush (3) to paint deep
textures such as pebbledash. Pour
some paint into a roller tray and dip the
brush in to load it. Scrub the paint onto
the wall using circular strokes to work it
well into the uneven surface.

Using a paint roller

A roller (4) will apply paint three times
faster than a brush. Use a long-pile
roller for heavy textures and a medium-
pile for lightly textured or smooth walls.
Rollers wear out very quickly on rough
walls, so have a spare sleeve handy.
Vary the angle of the stroke when using
a roller to ensure an even coverage,
and use a brush to cut into angles and
obstructions.
   A paint tray is difficult to use at the
top of a ladder unless you fit a tool
support, or, better still, erect a flat
platform from which to work.

Using a spray gun

Spraying is the quickest and most
efficient way to apply paint to a large
expanse of wall, but you will have to
mask all the parts you do not want to
paint, using newspaper and masking
tape, and erect plastic screening to
prevent overspray. Thin the paint by
about 10 per cent and set the spray gun
according to the manufacturer's
instructions to suit the particular paint.
It is advisable to wear a respirator.
   Hold the gun about 225mm (9in) away
from the wall and keep it moving with
even, parallel passes. Slightly overlap
each pass and try to keep the gun
pointing directly at the surface – tricky
while standing on a ladder. Trigger the
gun just before each pass and release it
at the end of the stroke.
   When spraying a large, blank wall,
paint it with vertical bands, overlapping
each band by 100mm (4in).
   Spray external corners by aiming the
gun directly at the apex so that paint
falls evenly on both surfaces (5). When
two walls meet at an internal angle,
treat each surface separately (6).

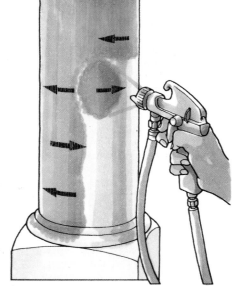

Spray-painting columns
Columns (such as those forming part of a front
door portico, for example) should be painted in a
series of overlapping vertical bands. Apply the
bands by running the spray gun from side to side
as you work down the column.

ROOFS: PITCHED ROOFS

Most pitched roofs were once built on site from individual lengths of timber, but to save time and materials, most builders now use prefabricated frames called trussed rafters. These are specifically designed to meet the loading requirements of a given house and, unlike traditional roofs, are not usually suitable for conversion because to remove any part of the structure can cause it to collapse.

Close-couple roof
A roof structure which has its rafters joined by joists. A variation is the collar roof where the joists (collars) are set at a higher level.

SEE ALSO
Details for:
Hipped roof 44-45
Roofing battens 46

Basic construction

The framework of an ordinary pitched roof is based on a triangle, the most rigid and economical form for a loadbearing structure. The weight of the roof covering is carried by the sloping members, the 'common rafters', which are set in opposing pairs whose heads meet against a central 'ridge board'. The lower ends, or feet, of the rafters are fixed to timber wall plates which are bedded on the exterior walls and distribute the weight uniformly.

To stop the roof's weight pushing the walls out, horizontal joists (ties) are fixed to the ends of each pair of rafters and the wall plates, forming a simple structure called a close-couple roof. The joists usually support the ceiling plaster; the rafters are also linked by roof-tiling battens.

TYPES OF PITCHED ROOF

Single roofs

Any roof, pitched or flat, with unbraced rafters – except for the pitched roof's joists – is called a single roof (**1**), and is only suitable for light coverings and short spans. For a wide span or heavy covering, the design would need unduly large roof timbers.

Double roofs

In double roofs, horizontal beams called purlins support the rafters (**2**), linking them either midway between foot and ridge or 2.5m (8ft) apart. This reduces the span of the rafters and allows light-weight timber to be used, 100 x 50mm (4 x 2in) being common. The cross section of a purlin depends to a great extent on the weight of the roof covering, but it usually exceeds that of the rafters; 200 x 50mm (8 x 2in) is normal. The ends of the purlins are supported on the brickwork of a gable wall or, in a hipped roof, by hip rafters.

To keep the size of the purlins to a minimum, struts are set in opposing pairs to brace them diagonally at every fourth or fifth pair of rafters. Struts transfer some weight back to the centre of the ceiling joists, which are supported there by a loadbearing dividing wall at right angles to them. The ends of the struts are jointed to a horizontal 'binder' fixed to the joists directly above the supporting wall.

Where ceiling joists are fairly lightweight and the span could make them sag, vertical timber 'hangers' are suspended from the top of every third and fourth rafter or from adjacent purlins at like intervals. At the bottom, hangers are fixed to a binder running at right angles across the joists.

Trussed roofs

Some traditional roofs embody trusses, rigid triangular frames that allow for a wider span, dispensing with loadbearing partition walls. Trusses carry the purlins, which in turn support the rafters and form a 'triple' or 'framed' roof. The trusses are spaced at 1.8m (6ft) or more, depending on the purlins' section or the roof covering's weight. As main bearers for the roof, they transmit weight to the exterior walls. Few trussed roofs can be converted and you should not try to cut into them.

'Trussed rafters' are now used in all new housing (**3**). Computer-designed for economy plus rigidity, the trusses are prefabricated of planed softwood 38mm (1½in) thick and up to 150mm (6in) wide, depending on roof loading. Each truss combines two common rafters, a joist and strut bracing in one frame; the members are butt-jointed and fixed with special nailed plate connectors.

The trusses are spaced a maximum 600mm (2ft) apart, linked horizontally with bracing members nailed to the struts. Such roofs are relatively lightweight, and are usually fixed to the walls with steel anchor straps to resist wind pressure.

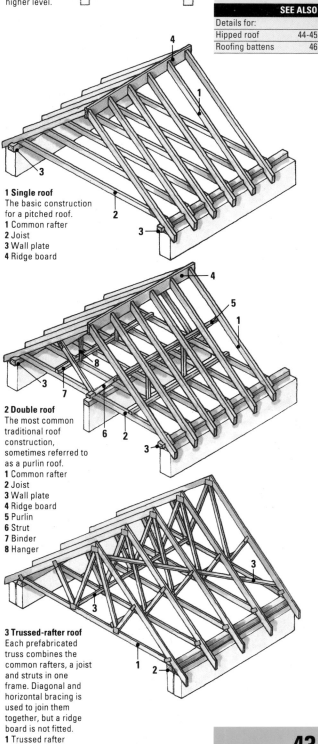

1 Single roof
The basic construction for a pitched roof.
1 Common rafter
2 Joist
3 Wall plate
4 Ridge board

2 Double roof
The most common traditional roof construction, sometimes referred to as a purlin roof.
1 Common rafter
2 Joist
3 Wall plate
4 Ridge board
5 Purlin
6 Strut
7 Binder
8 Hanger

3 Trussed-rafter roof
Each prefabricated truss combines the common rafters, a joist and struts in one frame. Diagonal and horizontal bracing is used to join them together, but a ridge board is not fitted.
1 Trussed rafter
2 Wall plate
3 Bracing

43

ROOF
CONSTRUCTION

Hips and valleys
All the components mentioned earlier are found in ordinary gable roofs. One with a hipped end or valleys has even more parts, but all serve to fulfil similar functions. Below, a double roof is used to illustrate the components.

RIDGE BOARD
COMMON RAFTER
PURLIN
CRIPPLE RAFTER
LAY BOARD
WALL PLATE
CROWN RAFTER
HIP RAFTER
JACK RAFTER

Hips and valleys

Eaves
That part of a roof where the rafters meet the exterior walls is known as the eaves. When rafters are cut flush with the walls, a fascia board is nailed horizontally across their ends to protect them and support the guttering (1).

1 Flush eaves

Open eaves
The ends of projecting rafters can be left exposed (2), with gutter-fixing brackets screwed either to their sides or to their top edges.

2 Open eaves

Closed eaves
Projecting rafters can also be clad with a fascia that is grooved to take a soffit board, enclosing the eaves (3). The soffit board can be at 90 degrees to the wall or slope with the rafters. If the loft is insulated, a roof with closed eaves must be ventilated with soffit vents.

3 Closed eaves

The verge
The verge is the sloping edge of the roof. It can end flush with the gable wall or project past it. With a flush verge, the end rafter fits inside the wall, but the roof covering extends over it (4). A projecting verge is constructed with the roof timbers extending beyond the wall to carry an external rafter with a 'barge board' fixed to it. There is often a soffit board behind the barge board to enclose a projecting verge (5).

ROOF-STRUCTURE PROBLEMS

A roof structure can fail as a result of timber decay caused by poor weatherproofing, condensation or insect attack. It can also suffer from overloading, especially if the timbers were inadequate in the first place. It is important to check that new roofing is not too heavy and to ensure that a window opening is braced properly. A sagging roof is often visible from street level, but it pays to inspect the roof structure closely from inside.

Inspection
The roof should be inspected annually to check that it is still weatherproof and that there is no woodworm infestation. If your loft has no natural lighting, buy a powerful torch or rig up a mains-powered extension lead with a caged lamp. In an unboarded attic, place planks across the joists to walk on.

Rot and infestation
Rot in roof timbers is a serious problem which should rectified by experts, and its cause should be identified and dealt with promptly. Rot is the result of damp conditions that encourage wood-rotting fungi to grow. Inspect the roof covering closely for loose and damaged slates or tiles in the vicinity of the rot; on a pitched roof water may be penetrating the covering at a higher level, so the leak may not be immediately obvious. If the rot is close to a gable wall you should suspect the flashing. Rot can also be the result of condensation. Better ventilation is the usual remedy.

If you employ contractors to treat the rot, it is better to have them make all the repairs. Their work is covered by a guarantee which may be invalidated if you attempt to deal with the cause yourself to save money.

Wood-boring beetle infestation should also be treated by professionals if it is serious. Severely infected wood may have to be replaced, and the whole structure will have to be sprayed.

Strengthening the roof
A sagging roof may not require bracing as long as the structure is sound, stable and weatherproof. Old houses with slightly sagging roof lines are often considered attractive, but consult a surveyor if you suspect a roof is weak.

The walls under the eaves should be inspected for bulging and checked with a plumb line. Bulging tends to occur where window openings are close to the eaves, making a wall relatively weak. However, bulging is sometimes caused by an inadequately braced roof structure that is spreading and pushing the walls outwards. If this proves to be the case, call in a builder or roofing contractor to do the repair work.

A lightly constructed roof can be strengthened by adding extra timbers. The exact method depends on the type of roof, its span, loading and condition. It may be possible to add sufficient bracing from inside the roof, provided the new timbers are not too large. If not, at least some of the roof covering will have to be stripped off.

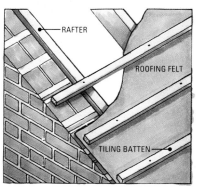

4 Flush verge

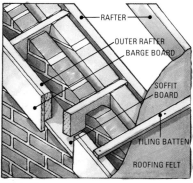

5 Projecting verge

PITCHED-ROOF
COVERINGS

SEE ALSO

Details for:	
Nails	46
Covering methods	46-47
Maintenance	48-49
Flashings	54-55

TYPES OF ROOFING

Roof coverings are manufactured by moulding clay or concrete into various profiles or by cutting natural materials such as slate into flat sheets.

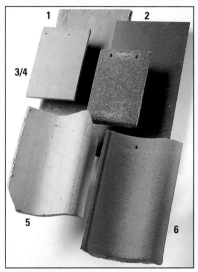

Selection of roof-covering materials
1 Natural slate. 2 Machine-made slate. 3 Plain tile (clay). 4 Plain tile (concrete). 5 Plain pantile (clay). 6 Interlocking pantile (concrete).

Coverings for domestic pitched roofs follow a long tradition, and despite the developments in new materials the older ones and the ways of using them have not changed radically.

Like most early building materials those used for roofing were generally of local origin, which led to a diversity of roof coverings, including tiles, slates and timber shingles. For centuries they were hand-made and had their own characteristics, visible in various regional styles.

During the last century the more durable roof coverings, such as tiles and slates, became more widely adopted.

Most roofing materials are laid across the roof in rows called courses so that the bottom of each overlaps the top of the one below. This means they are laid working from the eaves up the slope of the roof to the ridge.

Specially shaped tiles are used for capping the ridge or hips so as to weatherproof the junctions of the slopes. Where the covering meets a chimney or a wall it is protected with flashing, usually made of lead or mortar

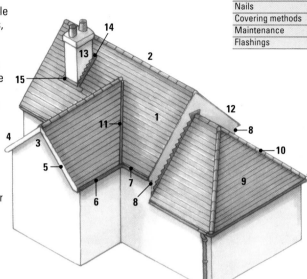

Roof-covering components for a pitched roof.
1 Tile or slate covering
2 Ridge tile
3 Gable end
4 Projecting verge
5 Barge board
6 Eaves
7 Fascia
8 Soffit
9 Hipped end
10 Hip tile
11 Valley
12 Flush verge
13 Lead stepped flashing
14 Back gutter
15 Apron

TYPICAL COVERINGS FOR PITCHED ROOFS							
Material	Common sizes	Finish	Colour	Fixing	Weight* kg/m² (lb/sq yd)	Minimum pitch in degrees	
TYPE OF COVERING							
SLATE	Split metamorphic sedimentary rock	Lengths: 300 to 600mm (1ft to 2ft) Widths: 180 to 350mm (7in to 1ft 2in)	Natural	Natural Blue Grey Green	Two nails	27.5 to 70 (51 to 129)	17½°
MACHINE-MADE SLATE	Fibre cement Asbestos cement (now rarely used)	Lengths: 400, 500, 600mm (1ft 4in, 1ft 8in, 2ft) Widths: 200, 250, 300mm (8in, 10in, 1ft)	Acrylic coating	Grey Blue/black Brown Russet Mottled	Two nails plus copper-disc rivet	18.5 to 22 (34 to 40)	20°
STONE	Split sandstone or limestone sedimentary rock Machine made	Random and as natural slate Lengths: 200 to 550mm (8in to 1ft 9½in) Widths: 100 to 500mm (4in to 1ft 8in)	Natural	Natural Yellow Grey Green	Two nails	90 (166) 84 to 110 (155 to 203)	20°
SHINGLES	Split or sawn red cedar	Length: 400mm (1ft 4in) Widths: 75 to 300mm (3in to 1ft)	Natural	Natural Brown Grey	Two nails	7 (13)	20°
PLAIN TILES	Hand-moulded or machine-moulded clay or machine-made concrete	Length: 265mm (10½in) Width: 165mm (6½in)	Sanded Smooth	Brown Red Grey Blue Green	Two nails or loose laid on nibs	76 to 87 (140 to 160)	40° clay 35° concrete
INTERLOCKING TILES	Hand-moulded or machine-moulded clay or machine-made concrete	Lengths: 380, 410, 430mm (1ft 3in, 1ft 4½in, 1ft 5in) Widths: 220, 330, 380mm (9in, 1ft 1in, 1ft 3in)	Sanded Smooth Glazed	Red Brown Grey Blue Green	Loose laid on nibs, nailed or clipped	40 to 57 (74 to 105) depending on profile	22½° clay 17½° to 30° concrete

*Approx

45

Tile clip

Nailed tile clip

Eave clip (flat)

Eave clip (contoured)

Verge clip (flat)

Copper rivet

TYPES OF ROOF COVERING

Double-lap coverings

Plain tiles, slates, stone 'slates' and wooden shingles are all double-lap coverings. They are basically flat – with the exception of plain tiles, which have a slight camber and nibs – and are laid with their side edges butting together, not overlapping. To prevent water penetrating the joints each course is lapped in part by the two courses above it. The joints are staggered, or 'broken jointed', on alternate courses like courses of bricks in a wall.

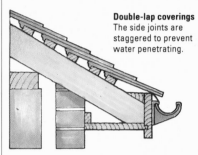
Double-lap coverings
The side joints are staggered to prevent water penetrating.

Single-lap coverings

Nearly all tiles of moulded clay and concrete are single-lap coverings, which means that each tile is profiled so as to interlock with the next one by means of a single lap on its side, and each course is laid with only a single lap at the head.

Early single-lap examples, such as clay pantiles, simply used the curved shape of the tile to form the overlap, but modern machine-made tiles of clay or concrete incorporate systems of grooves and water bars which prevent water penetrating the lap. These tiles have nibs at their top back edges which hook on to the battens. They are held in place by their own weight or are fixed with nails or clips.

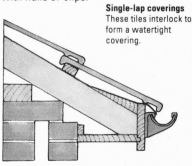
Single-lap coverings
These tiles interlock to form a watertight covering.

If you intend to make repairs yourself you will need some knowledge of the roof-covering system used on a common pitched roof. It will also help you if you have to commission contractors, either for repairs or reroofing work, as you will benefit from a better understanding of the work that is being carried out.

Underlay

To comply with current Building Regulations, new and re-covered pitched roofs must have a weather-resistant underlay of some kind.

This underlay, sometimes called sarking, should be a reinforced bituminous felt, Type 1F, or a suitable tear-resistant plastic material like polyethylene. These are sold in rolls to be cut to length as required.

The sheet material provides a barrier to any moisture that may penetrate the outer covering. It also improves the insulation value of the roof.

Like the tiles themselves, the sarking is laid horizontally, working upwards from the eaves, each strip overlapped by the one above it.

Battens

The roof covering is supported on sawn softwood battens which are nailed across the rafters, over the sarking (1). They are pretreated with a preserver. When the roof is close-boarded there should be vertical counter-battens under the horizontal battens (2) to provide some ventilation under the tiles and to allow any moisture to drain freely down the roof.

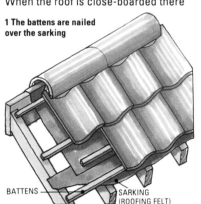
1 The battens are nailed over the sarking

BATTENS
SARKING (ROOFING FELT)

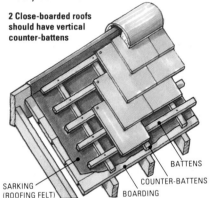
2 Close-boarded roofs should have vertical counter-battens

BATTENS
COUNTER-BATTENS
BOARDING
SARKING (ROOFING FELT)

Fixings

Most roof coverings are fixed with nails or clips. Slates and shingles are fixed individually with nails, normally two placed halfway up, though some are nailed at the top. Fibre-cement slates are centre-nailed with copper rivets to hold down the tails (see left).

Tiles have nibs that hook over the battens, keeping them in place, and some types need nothing more. Others are held with nails or clips.

The fixings are determined by the type and size of tile, the pitch of the roof and the exposure of the building.

● Black dot denotes that nail type and roof covering are compatible.

TYPE OF ROOF COVERING						
TYPE OF NAIL	Slate	Fibre-cement slate	Clay tiles	Concrete tiles	Shingles	Felt
COPPER	●	●	●	●	●	
ALUMINIUM-ALLOY	●		●	●	●	
SILICON-BRONZE	●	●	●	●	●	
GALVANIZED-STEEL					●	●
STAINLESS-STEEL	●	●	●	●	●	

GENERAL
CONSTRUCTION

ROOF SAFETY

Working on a roof can be hazardous, and if you feel insecure working at that height, you should hire a contractor. If you decide to do it yourself, do not use ladders alone to reach the roof; hire a sectional scaffold tower and scaffold board to provide a safe working platform complete with toe boards.

Roof coverings are fairly fragile and may not bear your weight – hire crawl boards or special roof ladders to gain access. A roof ladder should reach from the scaffold tower to the ridge of the roof and hook over the latter. Wheel the ladder up the slope and then turn it over to engage the hook (**1**).

Roof ladders are made with rails that keep the treads clear of the roof surface and spread the load (**2**), but if you think it necessary you can place additional padding of paper-stuffed or sand-filled sacks to help spread the load further.

Carry your tools in a special belt and only put them down within the roof-ladder framework. Make sure you bring every tool down from the roof when you have finished work.

1 Engage the hook of the ladder over the ridge

2 A roof ladder spreads the load

It is important for the overall appearance and performance of the roof that the covering is well finished at the verge, eaves and valley edges.

Verges

In the interest of neatness the verge is normally formed by first laying an undercloak of plain tile or slate bedded on the brickwork or – in the case of an overhanging verge – nailed to the timber frame. The roof covering is then bedded in mortar on top of the undercloak and finished flush. The verge of a slate or plain tiled roof is set to slope inwards slightly to prevent rainwater running down the walls, but single-lap tiles are laid flat.

Special dry-fixed verge tiles are available for use with single-lap concrete tiles and fibre-cement slates.

Eaves

The detail of the roof covering at the eaves depends on the type of covering. Plain tiling begins with a course of short under-tiles, nailed to a batten and projecting 38 to 50mm (1½ to 2in) over the gutter. The first course of whole tiles is laid with staggered joints over the under-tiles with their tail edges flush (**1**).

Some single-lap tiles, such as pantiles, also use an undercloak of plain tiles, the first course being bedded in mortar that fills up the hollow rolls. As an alternative there are special eaves tiles with blocked ends or overhangs.

Single-lap low-profile tiles are normally laid directly over, and supported by, the fascia board (**2**).

When the roof covering is natural slate, a double course is laid at the eaves. A course of short slates is nailed to the first batten and covered by a course of full slates with their tails flush and their joints staggered.

Three courses are used for fibre-cement slates to support the tail rivets used with this type of covering.

Valleys

Traditional double-lap roof coverings of plain tiles may embody special valley tiles, or, as with slate, may be formed into 'swept' or 'laced' valleys. The latter call for great skill and are expensive to make. Most valleys are formed as open gutters using sheet metal.

Single-lap roofing may use sheet valleys or special trough units.

An undercloak course projects slightly and gives a neat finish to the verge

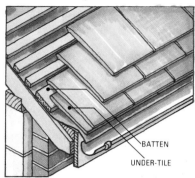

BATTEN
UNDER-TILE

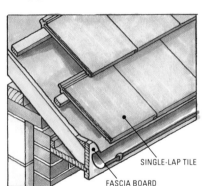

SINGLE-LAP TILE
FASCIA BOARD

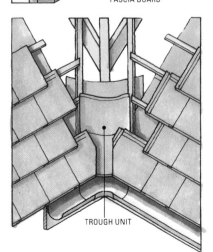

TROUGH UNIT

Verge detail at ridge
The ridge tile is set flush with the verge and filled with bedding mortar.

1 Plain tiles
The eaves under-tiles are nailed to a batten. The joints between them are covered with full tiles.

2 Single-lap low-profile tiles
Not all interlocking tiles need under-tiles at the eaves, but the fascia must support the eaves course at the correct angle.

Valley tiles
Modern roofs may have trough units instead of the traditional lead sheeting.

47

ROOF MAINTENANCE

The roof and upper parts of a building, such as chimneys and parapet walls, must be kept in sound condition if they are to remain weatherproof. Failure of the roof covering can cause expensive deterioration of the underlying timber structure, interior plaster and decorative finishes.

All roof coverings have a limited life, the length depending on the quality of materials used, the workmanship and exposure to severe weather. An average roof covering might be expected to give good service for 40 to 60 years, and some materials can last for 100 years or more, though some deterioration of the fixings and the flashings will take place. Reuse the old materials if you can.

Patch repairs may prove to be of only temporary value and can look unsightly. If patching becomes a recurrent chore it is time for the roof to be re-covered. This involves stripping off the original material and possibly reusing it, or perhaps replacing it with a new covering similar to the old. Major roof work is not something you should tackle yourself. A contractor will do it more quickly and will guarantee the work.

Reroofing work may qualify for a discretionary improvement grant from your local authority, so check with them before carrying out any work if you think you are eligible. You will not require planning permission unless you live in a listed building or a conservation area.

Inspecting the roof
The roofs of older houses are likely to show their age and should be checked at least once a year.

Start by taking a general look at the whole roof from ground level. Slipped or disjointed tiles or slates should be spotted easily against the regular lines of the undisturbed covering. The colour of any newly exposed and unweathered slate will also pinpoint a fault. Look at the ridge against the sky to check for misalignment and gaps in the mortar jointing. Follow this with a closer inspection through binoculars, checking the state of the flashings at abutments and around the chimney brickwork.

From inside an unlined roof you can spot chinks of daylight that indicate breaks in the covering. Use a torch to check the roof timbers for water stains; they may show as dark or white streaks. Trace the stain to find the source.

1 Pull out nails

2 Nail strip to batten

3 Fold strip over edge

Removing and replacing a slate

A slate may slip out of place because the nails have corroded or because the slate itself has broken. Whatever the cause, slipped or broken slates should be replaced as soon as possible before a high wind strips them off the roof.

Use a slater's ripper to remove the trapped part of a broken slate. Slip the tool under the slate and locate its hooked end over one of the fixing nails **(1)**, then pull down hard on the tool to extract or cut through the nail. Remove the second nail in the same way. Even where an aged slate has already slipped out completely you may have to remove the nails in the same way to allow the replacement slate to be inserted.

You will not be able to nail a new slate in place. Instead, use a plastic clip or cut a 25mm (1in) wide strip of lead or zinc to the length of the slate lap plus 25mm (1in). Attach the strip to the batten by driving a nail between the slates of the lower course **(2)**, then slide the new slate into position and turn back the end of the lead strip to secure it **(3)**.

Cutting natural slate

Cut from each edge

You may have to cut a second-hand slate to fit the gap in your roof. With a sharp point, mark out the required size on the back of the slate. Place the slate, bevelled side down, on a bench. Align the cutting line with the edge of the bench, then chop the slate with the edge of a bricklayer's trowel. Work from both edges towards the middle, using the edge of the bench as a guide. Drill nail holes or punch them out with a masonry nail – a punched hole leaves a recess for the head of a roofing nail.

Fibre-cement slates
Having scribed the lines, break a fibre-cement tile over a straightedge or cut it to size with a universal saw. If you saw asbestos-cement slates, wear a mask, keep the dust damped down and sweep it into a plastic bag for disposal. Make nail holes with a drill.

REPLACING A TILE

Individual tiles can be difficult to remove on two accounts: the retaining nibs on their back edges and their interlocking shape which holds them together.

Remove a broken plain tile by lifting it so that the nibs clear the batten on which they rest, then draw it out. This is easier if the overlapping tiles are first lifted with wooden wedges inserted at both sides of the tile that is to be removed **(1)**. If the tile is also nailed, try rocking it loose. If this fails you will have to break it out carefully. You may then have to use a slater's ripper to extract or cut any remaining nails.

Use a similar technique for a single-lap interlocking tiles, but in this case you will also have to wedge up the tile to the left of the one being removed **(2)**. If the tile has a deep profile you will have to ease up a number of surrounding tiles to achieve the required clearance.

If you are removing a tile to put in a roof vent, you can afford to smash it with a hammer, taking care not to damage any of the adjacent tiles. The remaining tiles should be easier to remove once the first is out.

1 Lift the overlapping tiles with wedges

2 Lift interlocking tiles above and to the left

CUTTING TILES

To cut tiles use an abrasive cutting disc in a power saw or hire an angle grinder for the purpose. Always wear protective goggles and a mask when cutting with a power tool.

To save money, use a tungsten-grit blade in a hacksaw frame to cut a few tiles. Trim the edges of a tile with pincers, but score the cutting line first with a tile cutter.

SHEET
ROOFING

REBEDDING RIDGE TILES

When the old lime mortar breaks down, a whole row of ridge tiles can be left with practically nothing but their weight holding them in place.

Lift off the ridge tiles and clear all the crumbling mortar from the roof and from the undersides of the tiles. Give the tiles a good soaking in water before starting to fix them.

Mix stiff mortar from 1 part cement : 3 parts sand. Load a bucket about half full and carry it on to the roof.

Dampen the top courses of roof tiles or slates, and lay a thick bed of mortar on each side of the ridge, following the line left behind by the old mortar (1). Lay mortar for one or two tiles at a time.

Press a ridge tile firmly into the mortar and use a trowel to slice off mortar that has squeezed out. Try not to smear any on the ridge tile.

Build up a bed of mortar to fill the hollow end of each ridge tile, inserting pieces of tile or slate to prevent the mortar slumping (2). Press the next tile in place, squeezing out enough mortar to fill the end joint flush. Build a similar mortar joint between ridge tiles and a wall or chimney stack.

1 Apply bands of bedding mortar on each side

2 Insert pieces of slate in joint bedding mortar

HALF-ROUND   HOG-BACK   ANGLE

Typical ridge tile shapes

The commonest sheet materials for roofing are made from fibre cement or rigid plastic (PVC). Aluminium, steel and corrugated bitumen sheeting are used for roofing, but not often for domestic work.

Sheet roofing is used mainly for outbuildings such as garages and garden sheds, and translucent plastic is used for lean-to extensions.

Consult your local Building Control Officer when considering a plastic roof to ensure that it complies with the fire regulations.

Corrugated-sheet roofing

Corrugated sheets are produced in standard profiles of 32mm (1¼in) and 75mm (3in) for plastic, and 75mm (3in) and 150mm (6in) for fibre cement.

When calculating the number of corrugated sheets you need, make an allowance for the side overlap. Small-profile sheets should overlap by at least two corrugations (1), while larger ones require an overlap of one only (2). The ends should overlap by at least 150mm (6in) in sheltered locations for roofs with pitches of about 22 degrees or more. For pitches of less than this, allow a 300mm (1ft) overlap.

Corrugated sheets are supported by purlins. They are of wood in most domestic buildings, though some system-built garages embody steel sections. Wood screws or drive screws fix the roofing to wooden purlins (3) and hooked bolts are used with metal ones (4). There are special plastic washers and caps for sealing the heads of the screws or bolts.

Cutting corrugated plastic
Mark the cutting lines on thin plastic sheet with a felt-tipped pen, then cut with a tenon saw. Support the sheeting between two boards on trestles and use the top board as a guide for your saw.

When cutting sheets to length, saw across the peaks of the corrugations with the saw held at a very shallow angle. Cut halfway through, then turn the sheet over and finish cutting from the other side.

When cutting to width, make the cut along the peak of a corrugation, again working with the saw at a shallow angle. Support a flexible sheet on two planks, one on each side of the line.

Cutting corrugated fibre cement
Lay the sheet on boards supported by trestles. As the material is fairly thick you will not need a top supporting board and you can saw the sheet without turning it over. Use a sheet saw or universal saw and wear a mask. Damp down dust from asbestos-fibre sheet and sweep it into a plastic bag. Seal the bag with adhesive tape ready for disposal. Damp newspaper under the trestles will help contain the dust.

1 Small-profile lap

2 Large-profile lap

3 Wood purlin fixing

4 Metal purlin fixing

Laying corrugated sheet

You can work from stepladders inside the structure or from above on boards, depending on the pitch of the roof.

Start at the eaves and work from left to right or vice versa.

Plastic sheet
Position the first sheet and drill oversized clearance holes for the fixing screws on the centre lines of the wooden purlins. Drill the holes in the crowns of the corrugations and space them at about every third or fourth corrugation. Never place fixings in the troughs of the sheeting.

Do not make holes where the next sheet will overlap the first. Instead, lay the second sheet with corrugations overlapping and drill through both sheets. Lay and fix the rest of the row of sheets in the same way.

Start the next row on the same side as the first one (1), overlapping the ends by at least 150mm (6in), and drill and fix through both layers.

Finally, fit the protective plastic caps over the fixing washers.

Laying fibre-cement sheet
Corrugated fibre-cement sheet is laid smooth side up. It is a fairly thick material, so cut some of the corners away to form mitres (2) in order to reduce the bulk of four thicknesses at the end laps. Draw the angle of the mitre between two points representing the length of the end lap and the width of the side lap.

Fix the sheets using the same method as for plastic sheet. For large-profile sheets use only two fixings to each purlin, placing them adjacent to the side laps. The end laps are fixed through both layers.

1 Overlap the ends

2 Mitre the corners

49

FLAT ROOFS

Timber-framed flat roofs are used for main roofs, rear extensions and outbuildings. Most have joists carrying stiff wooden decking, and these usually cross the shorter span, spaced at 400mm (1ft 4in), 450mm (1ft 6in) or 600mm (2ft) between centres. Herringbone or solid strutting is required for a span of more than 2.5m (8ft) to prevent the joists buckling. Their ends may be fixed to wall plates on loadbearing walls or, as on an extension, they are set in metal hangers or into the brickwork of the wall. Metal restraint straps tie the ends of the timbers down to the walls.

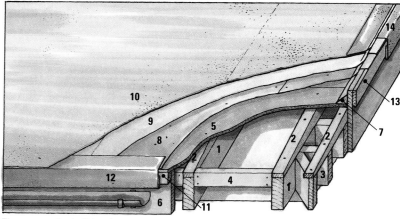

The components of a flat roof
1 Joists
2 Furring
3 Return-joist
4 Nogging
5 Decking
6 Fascia board
7 Angle fillet
8 First felt layer
9 Second felt layer
10 Third felt layer
11 Eaves drip batten
12 Felt eaves drip
13 Verge drip batten
14 Felt verge drip

Fall

The fall of a flat roof should be at least 1:80 for smooth surfaces like metal or plastic, and 1:60 for rougher materials. The fall is designed to shed water, but puddles may form if it is too shallow. The action of sun or frost can break down the roof covering and let the standing water through.

Sloping joists provide a fall, but this means that any ceiling below also slopes. To achieve a flat ceiling, tapered 'furrings' are nailed to the tops of the joists (1). Otherwise, joists may be set across the line of the fall with parallel furring pieces of decreasing thickness nailed to them (2) or tapered furrings are fixed across them (3). The latter provides better cross-ventilation.

Furring methods

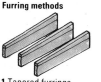

1 Tapered furrings fixed in line with joist.

2 Furrings of decreasing size fitted across fall.

3 Tapered furrings fixed across joists.

Decking

Chipboard or plywood decking is fixed to the joists to make a flat base for the final covering. Older flat roofs were decked with square-edged or tongue-and-groove boards which were laid with their joints running with the fall of the roof to shed water efficiently.

Plywood and chipboard panels, normally 18mm (¾in) thick, are laid with their long edges across the joists and their ends centred over supporting joists. Noggings may be fitted between the joists to give extra support to the long edges of the panels, depending on their thickness and joist spacing.

If you want a felted roof you might consider using prefelted chipboard. This material is laid with 3mm (⅛in) gaps between the boards to allow for thermal expansion, and fixed down with either nails or screws.

If you are unable to apply the final layer of felt sheeting immediately, you can make prefelted decking temporarily waterproof by filling the gaps between the boards with a cold-bonding mastic and then sealing the joints with 100mm (4in) wide roof-sealing tape. This is not a viable option for other types of decking.

Covering the deck
Whatever decking is used, it must be waterproofed with either mastic asphalt or roofing felt (see right).

In addition, the roof can be covered with a 12mm (½in) thick layer of pale-coloured chippings to reflect some of the sun's heat.

FLAT-ROOF COVERINGS

Bitumen-based coverings fall into two types: asphalt and bituminous felt. Felts are much better than they once were and are now generally used on domestic buildings instead of the lead, zinc or copper seen on older houses.

Mastic asphalt

This waterproof material, made from either natural or synthetic bitumen, weathers very well. It is melted in a cauldron and spread over the roof hot, to set in an impervious layer. Two layers are applied with a float to a combined thickness of 18mm (¾in) on a layer of sheathing felt covering the decking. Laying hot asphalt is a skilled job.

Roofing felts

These bitumen-impregnated sheet materials are applied in layers to produce 'built-up' roofing, bonded with hot or cold bitumen. Making such a roof with hot bitumen is a professional job.

Several felts are available, and the choice will affect a roof's cost and longevity. Traditional British Standard felts are classified by their reinforcing base material and finish, indicated by a number and letter. A colour strip identifies the base material. These felts are less tough than modern high-performance ones. The latter are based on glass tissue reinforced with polyester or polyester fabric, some with modified bitumen for greater flexibility.

Bonding felts

Plywood and chipboard decks need partial bonding of the first layer, using a perforated underlay either applied loose or bonded with bands of bitumen about 500mm (1ft 8in) wide around the edges. The first layer is then partially bonded by the hot bitumen penetrating through the holes. On solid timber, the first layer is secured with clout nails. Subsequent layers are fully bonded by applying hot or cold bitumen with a trowel or a notched spreader over the whole surface.

Butyl roofing

Butyl roofing is a single-ply flexible membrane that is nailed over a single layer of underfelt. It provides a strong, maintenance-free covering.

ABUTMENTS
AND PARAPETS

Leaks can occur wherever a flat roof abuts a house or parapet wall, so the roof covering is usually turned up the wall to form a 'skirting' which is tucked into the mortar bed of the brickwork, or covered by flashing.

Parapet walls are prone to damp, being exposed on both sides. Their top edges are usually finished with brick, stone or tile coping which should overhang the wall faces to throw off rainwater. Damp-proof courses of lead, bituminous felt or asphalt must be set in the mortar bedding beneath copings (1).

A parapet wall no more than 350mm (1ft 2in) high may have an asphalt skirting taken up the face and continued under the full width of the coping (2). Alternatively, the roof covering is taken up two courses of bricks only and is built into the wall to form a damp-proof course (3).

A flexible damp-proof course like bituminous felt or lead is often set in the bed joint before the roofing is laid and then dressed down to form a flashing over the skirting (4).

Cavity parapet walls also need a damp-proof course. Water penetrating from the roof side must be prevented from running down the cavity to damage the interior walls. The damp-proof course is stepped up from the inside leaf across the cavity to the outer leaf to form a cavity tray (5).

Cladding a parapet with cement rendering is not an entirely satisfactory solution as movement in the wall causes cracks that let in water.

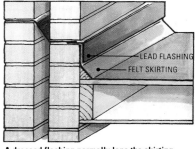
A dressed flashing normally laps the skirting

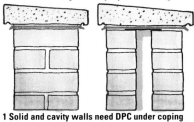
1 Solid and cavity walls need DPC under coping

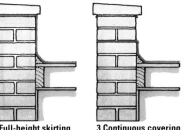
2 Full-height skirting   **3 Continuous covering**

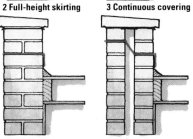
4 DPC flashing   **5 DPC cavity tray**

FITTING A NEW CAVITY TRAY

A cavity wall abutted by an extension roof needs a cavity tray to protect it from damp. Normally it would be built-in, but for existing buildings with new extensions, special trays – moulded units of polypropylene – can be inserted from outside by removing a course of bricks. A single tray, which equals two bricks in length, can also be used where a cavity is bridged by an extractor.

Inserting the tray
Remove three bricks, two courses above the proposed roof level. Do not let rubble fall into the cavity. On the cleaned bricks, lay a length of flashing wide enough to project 50mm (2in) into the wall and cover the roof skirting by 75mm (3in) when dressed down. Trap the flashing with the first tray unit, pushing it into one end of the opening (1). Use mortar to lay two bricks in the tray (2). Pack out the top joint with slate and fill it with mortar. Rake out a weep hole at the base of the middle joint to drain moisture from the cavity.

Cut out two more bricks, leaving a three-brick opening (3), roll out the flashing and insert a second tray. Join the trays with the clip provided, fitting it over the meeting ends to make a watertight joint (4). Lay two more bricks in the opening. Continue until the tray is the required length. You need only remove one brick at the end to make a two-brick opening for the last unit.

When the mortar is firm, point the new work to match the existing wall.

Moulded cavity tray Straight and angled sections are available from most builders' merchants.

1 Trap the flashing

2 Lay two bricks

3 Cut out two bricks

4 Join the trays

51

TYPES OF BITUMINOUS FELTS FOR FLAT ROOFS

Felt type British Standard Ref	Base	Surface Finish	Colour code	Weight kg/per roll	Properties and uses
BS 747 1B	Fibre	Sand	White	36kg (79lb)	Least expensive type. Relatively weak. Good for roofing outbuildings.
BS 747 1E	Fibre	Mineral	White	38kg (84lb)	
BS 747 3B	Glass fibre	Sand	Red	36kg (79lb)	Rot-proof, inexpensive, unsuitable for nailing. Good for 2- or 3-layer systems.
BS 747 3E	Glass fibre	Mineral	Red	28kg (62lb)	
BS 747 3G	Glass fibre	Grit underside Sand topside	Red	32kg (70lb)	Perforated first layer for partial bonding systems using hot bitumen.
HIGH-PERFORMANCE FELTS					
NO BS NUMBERS	Glass/polyester	Sand		36kg (79lb)	Rot-proof, tough, good weathering, can be nailed. Use for 2- or 3-layer systems.
	Glass/polyester	Mineral		28kg (62lb)	
BS 747 5U, 5B	Polyester	Sand	Blue	18.42kg (40.93lb)	More expensive than glass/polyester, but better performance. Use for 2- or 3-layer systems. Excellent for house extensions.
BS 747 5E	Polyester	Mineral	Blue	47kg (104lb)	
Elastomeric	Polyester	Sand		32kg (70lb)	Most expensive, but superior durability makes it long lasting and cost effective. Use for 2- or 3-layer systems. The best type for house extensions.
	Polyester	Mineral		40kg (88lb)	
	Polyester	Mineral		38kg (84lb)	

RENEWING A
FELT ROOF

Covering flat roofs, or stripping and re-covering them, should usually be left to professionals. A built-up felt system using hot bitumen or torching – using a gas-powered torch to soften bitumen-coated felt – is beyond the amateur. However, a competent person can confidently replace perished felt on a garage roof, using a cold-bitumen adhesive. The following example assumes a detached garage with a solid-timber decking covered with three layers of felt.

Replacing perished felt

Wait for dry weather, then strip the old felt. Pull out any clout nails and check the deck for distorted or rotten boards. Lift and replace unsound ones with new boards, using galvanized wire nails punched below the surface.

Cutting to stagger joints

For a three-layer build-up, start at one edge with a strip of felt about one-third of the roll width. The second layer starts with two-thirds width, the top layer with a full width. A two-layer roof starts with half a roll width, then with a full one. You can modify this to suit your roof and avoid having a strip that is too narrow at the other edge. If, for economy or ease of handling, you decide to use short lengths of felt, their ends should overlap by at least 100mm (4in), the lower piece always lapped by the higher one as you work up the slope.

First layer

Cut the strips for the first felt layer slightly longer than the slope of the roof and allow for an overlap of at least 50mm (2in) at the long edges. Cut the first narrow strip as described above.

Nail down the felt, using 18mm (¾in) clout nails 50mm (2in) apart down the centres of the laps and 150mm (6in) apart overall (1). Tuck and trim the felt into the corners of the verge upstand to achieve mitred butt joints. Trim the felt

flush at the eaves and verges, then form and fit the drip at the eaves (see right).

Second layer

Cut strips for the second layer, put the side piece in place, then roll it back halfway from the eaves end. Brush or trowel cold-bitumen adhesive on the felt below but not on the verge upstand. Re-lay the felt and press it down. Roll back the other half and repeat, ensuring that the adhesive is continuous.

Fold and tuck the felt into the corner of the verge upstand and trim it to a mitred butt joint. Turn it back from the verge, apply adhesive and press it into place against the upstand. Trim the end to butt against the edge of the felt eaves drip (see right) and trim the other edges flush with the verge. Place the next length overlapping the first by at least 50mm (2in) and again roll back each half in turn (2), applying the adhesive. Repeat across the roof, then cut and tuck the felt at the other verge corner.

Third layer

Cut and lay the mineral-felt top layer or 'capsheet' in the same way as the second layer, but this time lap the eaves drip and not the verge upstands (3).

Cut strips of mineral felt to form verge drips (see right), then nail and bond the strips into place around the side and rear edges.

Built-up felt system
Lap the edges of the felt strips and stagger the joints in alternate layers.

1 Nail first layer

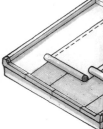

2 Bond second layer

3 Lap eaves drip

MAINTAINING A FLAT ROOF

Whatever the material used for covering your flat roof, it is sensible to carry out a routine inspection at least once a year.

Wear soft-soled shoes when you climb on to the roof, preferably soon after it has rained when you can see if there are standing puddles.

Remove old leaves, twigs and litter. Brush off any silt deposits that have built up, but take care you do not inadvertently lift the lapped edges of roofing felt. Note the positions of puddles because, although they may not present an immediate problem, you will

have a better idea of where to look if the roof springs a leak later on. You should also note any blisters or ripples on a roof covered with felt or asphalt. Check the overlapping joints on a felted roof to make sure they are still well-bonded.

Also make a close inspection of the vulnerable edges of the roof. Check the soundness of the covering at the verges and eaves and the flashings at abutments.

You should also inspect the condition of the gutters and downpipes, and remove any blockages.

MAKING DRIPS

Eaves

Cut 1m (3ft 3in) long strips from the length of a roll of felt. Calculate the width of the strips by measuring the depth of the drip batten and add 25mm (1in). Double this figure and add at least 100mm (4in).

Cut 50mm (2in) from one corner to enable the ends to be overlapped and make folds in the strips, using a straightedge. Nail the drip sections to the drip batten with galvanized clout nails, fold each strip back on itself and bond it over the first layer of felt (1).

Cutting the corners

Where the drip meets the verge, cut the corners to cover the end of the upstand (2). If necessary, make a paper pattern before cutting the felt. You may need an extra-wide strip to allow for a tall upstand. Fold the tabs and bond into place, except the end one, which is left free to be tucked into the verge drip.

Verge drips

Cut and fix the verge drips in place after the top layer of roofing. Cut the strips 1m (3ft 3in) long and calculate their width as with the eaves drip, but allow extra for the top edge and slope of the upstand. Working from the eaves, notch the ends of the strips where they overlap, as for eaves. Cut and fold the end of the first strip where it meets the eaves (3), nail the strip to the batten and bond the remainder in place (4).

At the rear corners, cut and fold the strip covering the side verge (5). Cover the rear verge last, cutting and folding the corners to lap the side pieces and finishing with neat mitres (6).

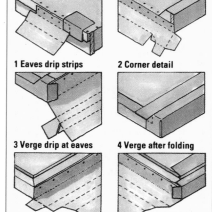

1 Eaves drip strips   **2 Corner detail**

3 Verge drip at eaves   **4 Verge after folding**

5 Verge corner   **6 Rear verge drip**

FAULT-FINDING

Damp patches
Damp patches on a ceiling are a clear sign that the roof needs attention, but the source of the problem is not always so obvious. If, however, the patches are close to a wall against which the roof abuts, you can be pretty sure the flashing has broken down.

Locating the leak
A leak anywhere else in the roof may be hard to find, as the water can run downhill from its entry point before dripping on to the ceiling. Measure the distance between the patch and the edges of the ceiling, then locate the point on the roof surface and work from there up the slope to find the source.

Splits and blisters
Splits and blisters on the smooth surface of an asphalt or bitumen felt covering may be obvious, but chippings on a covering tend to obliterate the cause of the leak. Use a blowtorch or hot-air paint stripper to soften the bitumen so you can scrape the chippings away. The surface must be smooth if it is to be patch-repaired.

Splits in the covering caused by movement of the substrate can be recognized by the lines they follow. Blisters formed by trapped moisture or air should be pressed to locate any weaknesses in the covering, which will show as moisture is expelled. These must be sealed with patches. They may be a result of moisture permeating the substrate from below and, heated by the sun, expanding under the covering. You can leave an undamaged blister for the time being, but deal with the cause as soon as possible.

Damp and condensation
Damp near a wall may be caused by porous brickwork above the flat roof, lack of pointing or damp-proof course, slipped or inadequate coping on parapet walls and/or a breakdown of flashings, which should be made good as required. Condensation also causes dampness, and can be a more serious problem. If warm, moist air permeates the ceiling, the vapour condenses under the cold roof and encourages rot in the structural timbers. In such a case, upgrade the ceiling with a vapour barrier and fit some type of ventilation. Otherwise, have the roof re-covered and include better insulation.

The best approach for repairing a flat roof depends on its general condition and age and the extent of the damage. If the surface of the covering has decayed, as may happen to some bitumen felts, it may be best to call in a contractor and have the roof re-covered.

Patch repairs
Such localized damage as splits and blisters can be patch-repaired with the aid of proprietary repair kits but, as their effectiveness is only as good as their adhesion to the background, take care when cleaning the surface. Kill any lichen or moss spores with fungicide or bleach before starting the repair work.

A patched roof, if visible from above, can be rather an eyesore. This can be corrected with a finishing coat of bitumen and chippings or reflective paint to unify the surface area. Work on a warm day, preferably after a spell of dry weather.

Dealing with splits
You can use most self-adhesive repair tapes to patch-repair splits in all types of roof coverings.

First remove any chippings (see left), then clean the split and its surrounding surface thoroughly. Fill a wide split with a mastic compound before taping. Apply the primer supplied over the area to be covered and leave it for an hour.

Where a short split has occurred along a joint in the board substrate, prepare the whole line of the joint for covering with tape.

Peel back the protective backing of the tape and apply it to the primed surface (1). If you are working on short splits cut the tape to length first. Otherwise work from the roll, unrolling the tape as you work along the repair.

Press it down firmly and, holding it in place with your foot, roll it out and tread it into place as you go, then cut it off at the end of the run. Go back and ensure that the edges are sealed (2).

1 Apply the tape

2 Press tape firmly

Dealing with blisters
Any blisters in asphalt or felted roofs should be left alone unless they have caused the covering to leak or they contain water.

To repair a blister in an asphalt roof, first heat the area with a blow torch or hot-air stripper and, when the asphalt is soft, try to press the blister flat with a block of wood. If water is present cut into the asphalt to open the blister up and let the moisture dry out. Apply gentle heat before pressing the asphalt back into place. Work mastic into the opening before closing it, then cover the repair with a patch of repair tape.

Make two intersecting cuts across a blister on a felted roof and peel back the covering. Heating the felt will make this easier. Dry and clean out the opening, apply bitumen adhesive and when it is tacky nail the covering back into place with galvanized clout nails (3).

Cover the repair with a patch of roofing felt, bonded on with the bitumen adhesive. Cut the patch so as to lap at least 75mm (3in) all round. Alternatively, you can use repair tape.

Treating the whole surface
A roof which has already been patch-repaired and is showing general signs of wear and tear can be given an extra lease of life by means of a liquid waterproofing treatment.

The treatment consists of a thick layer of cold-applied bitumen-based liquid waterproofer which can also be reinforced with an open-weave glass-fibre membrane.

First sweep the roof free of all dirt and litter, then treat the surface with a fungicide to kill off any traces of lichen and moss.

Following the manufacturer's instructions, apply the first coat of waterproofer with a brush or broom (4), then lay the glass-fibre fabric into the wet material and stipple it with a loaded brush. Overlap the edges of the fabric strips by at least 50mm (2in) and bed them down well with the waterproofer. Clean the brush with soapy water before the coating sets.

Let the first coat dry thoroughly before laying the second and allow that one to dry before applying the third and last coat. When the last coat becomes tacky cover it with fine chippings or clean sharp sand to provide it with a protective layer.

3 Nail cut edges
Use bitumen adhesive to glue a felt patch over the repair.

4 Brush on first coat

53

FLASHINGS

Flashings are used to weatherproof the junctions between a roof and the other parts of the building, which are usually at the abutments with walls and chimneys and where one roof meets another.

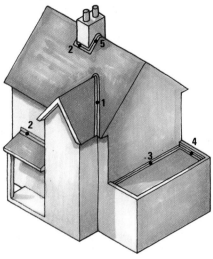

Where flashing is used
Typical types of flashing for pitched and flat roofs.
1 Valley
2 Apron
3 Wall abutment
4 Parapet abutment
5 Chimney abutment

Flashing materials

The most common flashing materials are lead, zinc, roofing felt and mortar fillets. Of all these, lead is by far the best because it weathers well, is easily worked – though shaping it is generally a craft skill – and can be applied to any situation and roof covering.

Zinc is a cheaper substitute for lead, but is not so long-lasting or so easy to work into shape. Consider using lead when zinc flashings need replacing – the extra cost buys a flashing that lasts considerably longer.

Bitumen felt may be used for flashings on felted roofs, but this material cannot be manipulated easily and is normally used for the more simple cover flashings that overlap the skirtings of felt roofs.

Mortar flashings, sometimes with inset cut tiles, are common on the pitched roofs of older houses. Although they tend to shrink and cause problems later, mortar flashings are still used as they are cheap and easy to apply.

FLASHING CONSTRUCTION

The design of a given flashing is determined by the particular details at the junction and to some extent by the materials used. Typical situations and methods are described here, using lead for the flashings.

Abutments

A flashing is used to seal the joints between the sloping edge of a pitched roof and abutting walls. The type of flashing is determined by the pitch and the nature of the roof covering.

Double-lap flashing

Slate or plain-tiled roofs with a pitch of 30 degrees or more normally use soakers and a cover flashing. Soakers are lead or zinc pieces, equal in length to a tile's overlap, folded at right angles lengthways. The part that lies on the tiles should be at least 100mm (4in) wide and the upstand 75mm (3in). The back edge turns down over the tile's top edge, so add 12 to 25mm (½ to 1in) to the length of the soaker. A soaker is laid over the end tile or slate as a course is laid. The upstand lies flat against the brickwork and is lapped by a stepped flashing dressed down over it. The flashing's top edges are turned into the bed joints, held by lead wedges and pointed with mortar (1).

Single-lap flashing

Contoured single-lap tiles can be treated at abutments with a one-piece flashing. The lead is tucked into the brick wall using the stepped method, and dressed down over the tile. The amount of overlap depends on tile contour and roof pitch. On a shallow pitch it should be at least 150mm (6in). The lead is dressed to the tile's shape and the step at each course, and its free edge is carried over the nearest raised tile contour (2).

Valley flashing

Some tiled roofs have valley tiles that take the tiling into the angle, but most tiled and slated roofs have metal valley flashings made by laying a lead lining on boarding that runs from eaves to ridge, following the angle of the valley. The lead is dressed over wood fillets nailed to the boarding to form an upstand (3). Where two valleys meet at the ridge, a lead saddle is formed. The edges of the tiles or slates are cut to follow the angle of the valley and to leave a gap of no less than 100mm (4in) between them.

Slate coverings should overhang the supporting valley fillet by 50mm (2in) and contoured tiles should be bedded in mortar and finished flush with the edge of the tiles to form a watertight gutter.

Apron flashing

The head of a lean-to roof is weatherproofed with a lead apron flashing, its top edge pointed into a mortar joint two courses above the roof. The lead is dressed down on to the roof and overlaps the the roof covering by 150mm (6in) or more.

Special moulded flashing units are available for use with corrugated-sheet roofing of plastic or fibre cement. These are shaped to fit the contour of the roofing and have flat hinged upstands which can be adapted to fit any roof slope. The upstand is lapped with a conventional flashing or is sealed with self-adhesive tape.

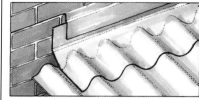

Moulded apron flashing for corrugated roofing

Chimney flashing

The flashing where a roof meets the side of a chimney is similar to that at an abutment, but there are also junctions at the front and back of the chimney.

An apron flashing is fitted in front. The upstand is returned on to the sides of the stack and its top edge is set in a joint in the brickwork. The apron, extending beyond the chimney sides, is dressed to the tile contour.

Single-lap or double-lap stepped flashings are fitted to the side of the chimney. At the back there is a timber-supported gutter. The front edge of the lead is turned up the brick face; its ends are folded over the side flashings. A separate cover flashing is dressed over the upstand at the rear of the chimney. The back of the gutter follows the roof slope and is lapped by the tiles, which are fitted last.

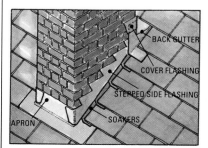

Chimney flashing for a slate roof

1 Double-lap flashing

2 Single-lap flashing

3 Valley flashing

There are many problems associated with flashings, usually as a result of the flashing material corroding or because the joints between different materials fail due to erosion or thermal movement.

A perished flashing should be stripped out and replaced. If this requires craft skills, the work should be done by a specialist contractor, but in many cases leaks are caused by shrinkage cracks which you can repair with self-adhesive flashing materials.

Caulking compound

Cement fillets often shrink away from wall abutments. If the fillets are otherwise sound, fill the gap with a gun-applied flexible caulking compound. Choose a colour that matches the fillet. Brush the surfaces to remove any loose material before injecting the compound.

Flashing tape

Prepare the surfaces by removing all loose and organic material. A broken or crumbling cement fillet should be made good with mortar.

Ensure that the surfaces are dry and if necessary apply a primer – it is supplied with some tapes – about one hour before using the tape (1).

Cut the tape to length and peel away the protective backing as you press the tape into place. Work over the surface with a cloth pad, applying firm pressure to exclude any air trapped beneath the tape (2).

1 Apply a primer with a 50mm (2in) paintbrush

2 Press tape with a pad to exclude air bubbles

Repointing flashing

Metal flashings which are tucked into brickwork may have worked loose where the old mortar is badly weathered. If the flashing is otherwise sound rake out the mortar joint, tuck the lead or zinc into it and wedge it there with rolled strips of lead spaced about 500mm (20in) apart. Then repoint the joint. While you have the roof ladders and scaffolding in place, rake out and repoint all of the mortar joints if they are in poor condition.

Rake out joint and repoint with fresh mortar

Patching lead

Lead will not readily corrode but splits can occur in it where it has buckled through expansion and contraction over the years. Flashing tape can be used but it is possible to patch lead by soldering or, for a more substantial repair, cutting away a weak or damaged portion and joining on a new piece by lead 'burning' or welding. This is not a job you can easily do yourself and it should be handed over to a professional. It should be done only when there is no risk of fire and it is more economical to have the old lead repaired than to have a new flashing fitted.

SEALED GLAZED ROOFS

Traditional porches, greenhouses and timber-framed conservatories all tend to suffer from leaks caused by a breakdown of the seal between the glass and glazing bars. Minor leaks should be dealt with promptly because trapped moisture can lead to timber decay and expensive repairs.

Using aluminium tape
You can waterproof glazing bars with self-adhesive aluminium tape simply cut to length and pressed in place.

Clean out old putty from both sides of the glazing bars, let the wood dry out and apply wood primer or linseed oil. When the primer is dry fill the rebates with putty or mastic.

The tape must be wide enough to cover each glazing bar and lap the glass on each side by about 18mm (¾in).

Start at the eaves and work up the roof, moulding the tape to each glazing bar and excluding air bubbles.

At a step in the glass, cut the tape and make an overlap. Mould the cut end over the stepped edge, then start a new length, lapping the stuck-down end by 50mm (2in).

At the ridge you may cut the tape to butt against the framework or lap on to it. Cover the ends with tape applied horizontally. Where a lean-to roof has an apron flashing tuck the tape under it.

Self-adhesive aluminium tape can be painted to match the woodwork or left its natural colour.

Mould the tape over the glazing bar

Clear tape
Make a temporary repair to cracked glass with clear self-adhesive waterproofing tape.

Clean the glass and apply the tape over the crack on the outside. It will make an almost invisible repair if applied promptly.

You can also use this tape for sealing the overlap on translucent corrugated-plastic roofing.

GUTTERING

Guttering collects rainwater that runs down a roof and leads it to a downpipe through which it discharges into a drain. Good rainwater disposal is vital in preventing damp developing in the fabric of a house.

Roof drainage

The size and layout of a roof drainage system should enable it to discharge efficiently all the water from a given roof area. Manufacturers of rainwater goods usually specify the maximum area for a given size and profile of gutter based on a rainfall rate of 75mm/hr (3in/hr). If you need to replace an old gutter, make sure you install one of the same size or perhaps slightly larger.

Very simple alterations can affect the performance of a drainage system quite dramatically. A system with a central downpipe, for example, can serve double the roof area of one with an end outlet. Conversely, a right-angle bend near the outlet can reduce the flow capacity by about 20 per cent.

In practice, unless you are working on an extension or a new garage, the positions of drains and downpipes are probably already fixed.

Gutter sizes
Sizes are generally specified by their overall width in cross-section and sometimes by their depth as well.

TYPICAL PROFILES AND SIZES OF GUTTERING

Half-round	Ogee (OG)	Moulded (OG)	Box
75mm (3in)			
100mm (4in)	100mm (4in)	100 x 75mm (4 x 3in)	100 x 75mm (4 x 3in)
112mm (4½in)	112mm (4½in)		
125mm (5in)	125mm (5in)	125 x 100mm (5 x 4in)	125 x 100mm (5 x 4in)
150mm (6in)	150 x 100mm (6 x 4in)	150 x 100mm (6 x 4in)	

EAVES-GUTTER SYSTEMS

Eaves-gutter systems are fabricated in cast iron, cast-aluminium, rolled-sheet aluminium, asbestos cement and a rigid uPVC plastic. With the exception of roll-formed aluminium types, the systems are made up from basic lengths of gutter and downpipes with a range of fittings (see diagram).

Moulded or cast gutters have a socket at one end into which the plain spigot end of the next section is jointed.

Gutters that are not symmetrical in section, such as an OG, require components that are left-handed or right-handed, and you will have to note this when ordering replacement parts.

Traditional cast-iron and modern aluminium guttering may be compatible should you wish to extend your system or renew part of it, but check carefully before purchasing. Patented plastic drainage systems, though superficially similar in style, are not always interchangeable.

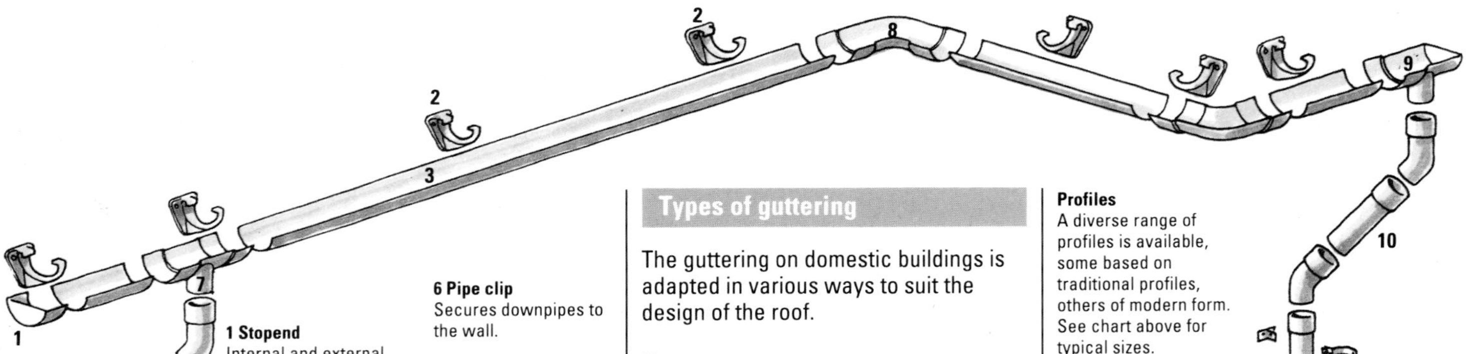

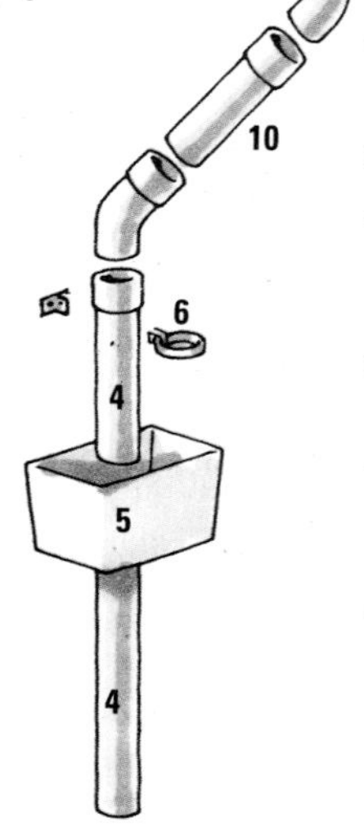

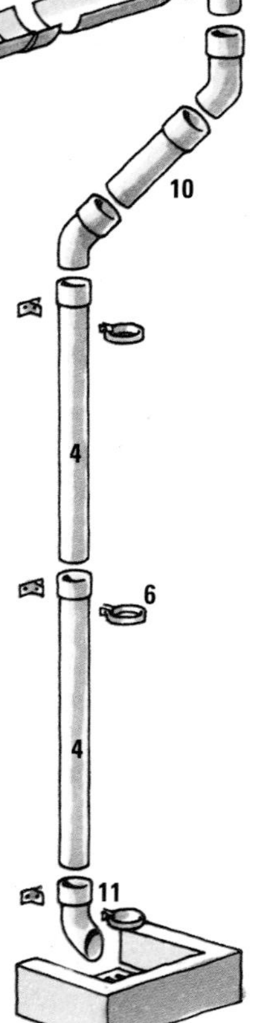

Types of guttering

The guttering on domestic buildings is adapted in various ways to suit the design of the roof.

Eaves gutters
Gutters that are fixed to fascia boards along the eaves of the roof are the most common form of guttering. They are made in many materials and a number of designs (see right).

Parapet gutters
Parapet gutters are generally found in older houses and may serve a flat or pitched roof set between two parapet walls. This type is generally purpose-made as part of the original structure of the roof and is usually covered with a metal or bituminous roofing material.

Valley gutters
Valley gutters are a form of flashing used at the junctions between sloping roofs. They are not gutter systems in themselves but they direct the rainwater into eaves or parapet gutters.

1 Stopend
Internal and external fittings for socketed or non-socketed types.

2 Gutter brackets
Normally screwed to fascia board, but some are fixed to rafter-bracket arms.

3 Guttering
In various profiles and lengths of 1.8 to 3m (6 to 10ft) with socket at one end or spigots at both ends.

4 Downpipe
Available in 1.8 to 3m (6 to 10ft) lengths. Metal types may have integral fixing lugs.

5 Hopper head
May be used as part of downpipe system to receive waste pipes from another source.

6 Pipe clip
Secures downpipes to the wall.

7 Running outlet
May have double or single sockets.

8 Gutter angle
Available in 90-, 120- and 135-degree angles in most systems for turning corners.

9 Stopend outlet
Used with downpipe at an end.

10 Offset
Used on guttering fitted to overhanging eaves. Available in standard projections or can be made up with special offset bends and a length of downpipe.

11 Shoe
Throws water clear of a wall into open gulley.

Profiles
A diverse range of profiles is available, some based on traditional profiles, others of modern form. See chart above for typical sizes.

Plain half-round

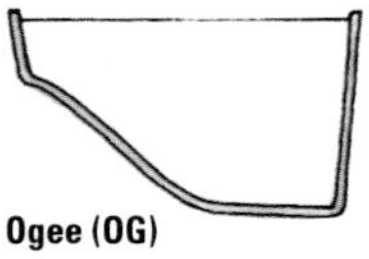

Ogee (OG)

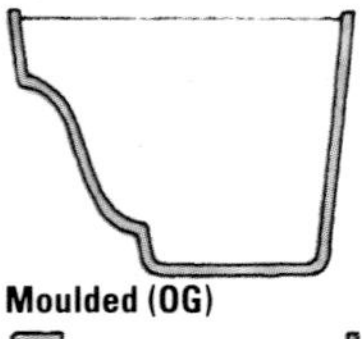

Moulded (OG)

Box

SEE ALSO
Details for:
Ladder stay   77

Cast-iron, cast-aluminium and asbestos-cement guttering are all rigid and may support a ladder, but it is much safer to use a ladder stay. Never prop a ladder against either plastic or roll-formed aluminium gutters.

Inspect and clean out the interior of gutters regularly. Gutters concentrate the dirt, and sometimes sand washed down from the tiles by the rain. This builds up quickly if the flow of water is restricted by leaves or twigs. Birds' nests also can effectively block the guttering or downpipes.

The weight of standing water can distort plastic guttering, and if a blockage causes the gutter to overflow, it leads to damp soaking through the wall below.

Removing debris

First block the gutter outlet with rag. With a shaped piece of plastic laminate scrape the silt into a heap, scoop it out of the gutter with a garden trowel and deposit it in a bucket hung from the ladder. Sweep the gutter clean with a stiff hand brush. Remove the rag and flush the gutter down with a bucket of water. Fit a wire or plastic 'balloon' in the end of the downpipe to prevent birds' nests or debris blocking the pipe.

Snow and ice

Plastic guttering can be badly distorted and even broken by snow and ice building up in it. Dislodge the build-up with a broom from an upstairs window if you can reach it safely. Otherwise, climb a ladder to remove it.

If snow and ice become a regular seasonal problem, fit a snow board made from 75 x 25mm (3 x 1in) planed softwood treated with a wood preserver and painted. Design it to stand about 25mm (1in) above the eaves tiles, using 25mm x 6mm (1 x ¼in) steel straps bent as required.

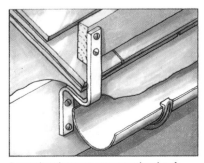

A snow board protects gutters or glazed roofs

GUTTERING MATERIALS

Cast iron

The cast-iron rainwater systems common on old houses are mostly of the OG type. They are fixed to the fascia board with short mushroom-headed screws that pass through the back of the gutter above the water line.

Each 1.8m (6ft) standard length of the guttering has a socket end into which the plain spigot end of the next piece fits (1). Short bolts secure the joint and a bedding of putty forms a seal when they are tightened (2).

1 A standard gutter has a socket at one end

2 The joint is sealed with putty then bolted

Cast iron is heavy and brittle, and installing or dismantling such a system needs two people. The iron can be cut with a hacksaw and drilled with twist drills in a power tool.

The guttering needs regular painting, and a bituminous paint applied inside helps to preserve the metal. If it is left unprotected it will rust, usually along the back edge, around the screws. Badly rusted guttering should be replaced, as it is likely to collapse.

Cast aluminium

Cast-aluminium guttering comes in a wide range of profiles. It is assembled in a similar way to cast-iron guttering with bolted joints, but a flexible mastic is used instead of putty to make the seals. The guttering may be fitted to the fascia with screws through the back, or with gutter brackets. The fixings should be sherardized or plated with zinc or cadmium. Cast aluminium is about one-third the weight of cast iron and can be left unpainted. However, in some situations it will corrode, and if it is used as part of a cast-iron system all the aluminium surfaces must be protected with zinc phosphate or with bituminous paint. It can be worked with ordinary metal-working tools.

Fibre cement

Fibre-cement guttering is of the socket-and-spigot type. The joints are secured with galvanized-iron bolts and the seals are made with a mastic jointing compound. The guttering is produced in half-round profiles in a range of several sizes. It is fixed to the house fascia board with gutter brackets, and these should not be spaced more than 900m (3ft) apart.

Fibre cement has good weathering properties and does not need to be painted. It can be cut with a hacksaw and drilled normally.

Rolled-sheet aluminium

Rolled-sheet aluminium guttering is a moulded lightweight OG system made from thin, prepainted flat-sheet aluminium which is roll-formed to the gutter shape by a portable machine. This is done on site by the suppliers, and continuous lengths are made to measure. The end stops and angles are supplied as separate items, crimped to the ends of the gutter sections. Outlets are formed by punching holes in the bottom. Simple metal fixing brackets are clipped to the front and back edges of the guttering and fixed to the fascia with drive screws. The system needs no maintenance, but can be painted.

Unplasticized PVC

Unplasticized PVC (uPVC) is now the most widely used guttering, both for new buildings and for replacing old systems. In many profiles and sizes, it is self-coloured in brown, black, grey and white. It needs no painting.

Most uPVC systems use clip-fastened joints with synthetic-rubber gaskets to form the seals. Some use a solvent cement to weld the joints. Downpipes may be a push fit, and are solvent-welded or sealed with an O-ring. The guttering is usually supported by brackets, but some systems employ screws to attach outlets and corner fittings for easier installation.

Cast-iron OG gutter

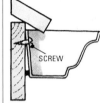

Cast aluminium

Gutter brackets

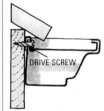

Sheet aluminium

Plastic guttering

57

FITTING NEW GUTTERING & DOWNPIPE

SEE ALSO
Details for:
Fascia board   44

When your old gutter system reaches the end of its useful life you should replace it. Try to do so with a system in the same style or, at least, one that complements the style of your house. If you plan to install the guttering yourself a plastic system is probably the best choice, being the easiest to handle.

Installing guttering

Measure round the base of the house to find the total length of gutter needed, and note the number and type of fittings to be ordered.

Use a plumb line to mark the position of the gutter outlet – directly over the existing drain – on the fascia board (1). Screw the outlet or its support bracket – depending on the system – to the fascia, no more than 50mm (2in) below the tile level (2).

Fix a gutter bracket at the opposite end of the run, close to the top of the fascia board. This is to provide a fall of at least 25mm (1in) in 15m (50ft) (3). Run a taut string between the bracket and outlet and fix the rest of the brackets to follow the slope of the string, spaced no more than 1m (3ft 3in) apart (4). There should be a bracket at every joint unless the system uses screw-fixed outlets and angles (5).

Fitting the gutter

6 Clip the guttering into the brackets

Tuck the back edge of a length of gutter under the roofing felt and into the rear lips of the brackets. Attach the front of each bracket in turn (6).

Fit the second length in the same way, with its spigot end pushed into the socket of the first length. You will need to apply pressure to compress the rubber seal. Leave a 6mm (¼in) gap between the end of the spigot and the shoulder of the socket to allow for expansion.

Cutting the gutter
Cut the gutter squarely with a hacksaw. You can snap a clip or bracket over it first to provide rigidity and guide the saw. Smooth rough edges with a file. Some systems require notches for the clips in the gutter's front and back edges; these can be made with a file.

7 Use an adaptor to join different systems

Connecting to existing guttering
Renewing guttering on a terraced house may mean joining your system to your neighbour's. There are left-hand and right-hand adaptors for this purpose.

Remove your old guttering to the nearest joint between the houses, bolt the adaptor on, sealing the joint with mastic, and fix the new plastic gutter with the clip provided (7).

Fitting the downpipe

Work downward from the gutter outlet. If the eaves overhang you will need to fit an offset.

Fit a clip to the top of a length of downpipe. Hold it against the wall and measure the distance from its centre to a plumb line dropped through the outlet's centre (8). You may find an offset to fit, but you will probably have to make one up with offset bends and short lengths of pipe. Use a solvent cement and assemble it on a table so that the bends lie in the same plane.

Fit the offset to the outlet spigot and the pipe to the offset. Adjust the pipe so that the clip's back plate falls on a mortar joint (9). Trim the offset spigot if necessary.

Mark the fixings, drill and plug the wall and fix the pipe and clip with plated round-head screws.

Mark and fix the lower lengths of pipe in the same way, with a 6mm (¼in) expansion gap between each pipe and the socket shoulder. Fit extra clips at the centre of pipes longer than 2m (6ft 6in).

Cut the bottom pipe to length and fit a shoe in the same way (10). If the pipe is jointed into a gulley trap (11) or a drain socket you may have to work upwards from the bottom.

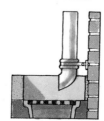

8 Drop a plumb line   9 Fit an offset   10 Finish with a shoe   11 Fit into gulley

Marking out new guttering
1 Mark fascia board.
2 Fix outlet or bracket.
3 Fix end bracket.
4 Slope the brackets.
5 Support the joints.

REPAIRING GUTTERS AND DOWNPIPES

It is always better to replace a damaged part of a gutter system than to repair it, but a neat repair that stays watertight can tide you over.

Mending a crack
All types of chipped or cracked guttering or downpipes can be repaired with a sprayable asphalt mastic.

Brush off loose material and spray on one or two coats from a distance of 300 to 400mm (12 to 16in) to seal the leak. Mask off the area around the repair. For large breaks, apply a coat of mastic and cover with a patch of fine glass-fibre matting or gauze. Spray on a coating of mastic to complete the repair.

Using epoxy putty
You can build up the chipped edge of a cast-iron gutter with epoxy putty. Tape a waxed or polyethylene-lined piece of card across the outside of the broken edge and bend it to follow the contour of the gutter. Mix the two-part putty, fill the gap with it and remove the cardboard 'former' when the putty sets.

Approximately a quarter of the heat lost from an average house goes through the roof, so preventing this should be one of your priorities when it comes to insulating your home. Provided that you are able to gain access to your loft space, reducing heat loss through the roof is just a matter of laying insulating material between the joists, which is cheap, quick and effective. If you want to use your attic, insulating the sloping surface of the roof is a quite straightforward alternative.

Treating a flat roof

A flat roof – perhaps on an extension – may also need insulating, but the only practical remedy for most house-holders is to apply a layer of insulation to the ceiling surface. It is not a particularly difficult task, providing the area is not too large, but you will have to relocate lighting and take into consideration any features, such as windows or fitted cupboards, that extend to the ceiling. Fixing ceiling tiles is an alternative, but their insulation value is minimal.

Preparing the loft

On inspection, you may find that your roof space has existing but inadequate insulation – at one time even 25mm (1in) of insulation was considered to be acceptable. It is worth installing extra material to bring the insulation up to the recommended thickness of 150mm (6in). Check roof timbers for woodworm or signs of rot, so they can be treated first. Make sure that all the electrical wiring is sound, and lift it clear so that you can lay insulation beneath it.

The plaster or plasterboard ceiling below will not support your weight. You therefore need to lay a plank or two, or a chipboard panel, across the joists so you can move about safely.

If there is no permanent lighting in the loft, rig up an inspection lamp on an extension lead and move it wherever it is needed – or hang the lamp high up to provide an overall light.

Most attics are very dusty, so wear old clothes and a gauze face mask. It is also wisest to wear protective gloves, especially if you're handling glass-fibre batts or blanket insulation, which may irritate sensitive skin.

TYPES OF INSULATION

There's a wide range of insulating materials available, so it is important to check the recommended types with your local authority before applying for a grant.

Blanket insulation

Blanket insulation, which is made from glass fibre, mineral fibre or rock fibre, is widely available in the form of rolls that fit snugly between the joists. The same material, cut to shorter lengths, is also sold as 'batts'. A minimum thickness of 150mm (6in) is recommended for loft insulation. Blanket insulation may be unbacked, or paper-backed to improve its tear-resistance, or it may have a foil backing that serves as a vapour barrier (see below).

The unbacked type is normally used for laying on the loft floor. The blankets are usually either 100mm (4in) or 150mm (6in) thick. Some 100mm blankets can be split into two for topping up existing insulation. The rolls are typically 6 to 8m (20 to 25ft) long and 400mm (1ft 4in) wide in order to be suitable for standard joist spacing. For wider-than-usual joist spacing, choose a roll 600mm (2ft) wide.

If you want to fit blanket insulation to the sloping part of a roof, buy it with a lip of backing along each side to staple to the rafters. Both mineral fibre and glass fibre are non-flammable, and are proofed against damp, rot and vermin.

Loose-fill insulation

Loose-fill insulation in pellet or granular form is poured between the joists, up to the recommended depth of 150mm. This will inevitably bury some joists, but you can nail strips of wood to the tops of the joists that support walkway boarding.

Exfoliated vermiculite, made from the mineral mica, is the most common form of loose-fill insulation – but other types, such as mineral wool, cork granules or cellulose fibre (made from recycled paper), may be available. Loose-fill is supplied in 10 or 20kg (22 or 44lb) bags. A 10kg (22lb) bag covers about 1.6sq m (17sq ft) to a depth of 150mm (6in).

It's not advisable to use loose-fill in a draughty, exposed loft, as high winds can cause it to blow about. On the other hand, it is convenient to use if the joists are irregularly spaced.

Blown-fibre insulation

Fibrous inter-joist insulation is blown through a large hose by a professional contractor. It may not be suitable for a house in a windy location, but seek a contractor's advice. An even depth of at least 150mm (6in) is required.

Rigid and semi-rigid sheet insulation

Sheet insulation, such as semi-rigid batts of glass fibre or mineral fibre, can be fixed between the rafters. They tend to perform better than lightweight rolls, so relatively thin sheets can be used, especially if covered with plasterboard. However, it pays to install the thickest insulation possible, allowing sufficient ventilation between it and the roof tiles or slates to avoid condensation.

VAPOUR BARRIERS

Installing roof insulation has the effect of making the uninsulated parts of the house colder than before, so increasing the risk of condensation either on or within the structure itself. In time, this could reduce the effectiveness of the insulation – and also promote a serious outbreak of dry rot in the roof timbers.

One way to prevent this happening is to provide adequate ventilation for the areas which are outside the insulation. Another solution is to install a vapour barrier on the warm (inner) side of the insulation to prevent moisture-laden air passing through. The vapour barrier is usually a plastic or metal-foil sheet, and is sometimes supplied along with the insulation. It is vital that the barrier is continuous and undamaged; otherwise its effectiveness is greatly reduced.

SEE ALSO	
Details for:	
Condensation	10
Dry and wet rot	15
Roof ventilation	71-72

ESTIMATING FOR BLANKET INSULATION
150 X 400mm wide rolls

Approx. loft area		
Square metres	Square feet	No. of rolls
30	332	17
34	366	20
38	409	22
42	452	25
46	495	27
50	538	29
54	581	32
58	624	34
62	667	36
66	710	39
70	753	41
74	796	43

Allows for average joist widths of 50mm (2in)

● **Ventilating the loft**
Laying insulation between the joists increases the risk of condensation in an unheated roof space – but provided there are adequate vents or gaps at the eaves, there will be enough air circulating to keep the loft dry.

59

INSULATING A LOFT

Laying blanket insulation

Before starting to lay blanket insulation, seal gaps around pipes, vents or wiring entering the loft with flexible mastic.

Remove the blanket wrapping in the loft (the insulation is compressed for storage and transportation, but swells to its true thickness when released), and begin by placing one end of a roll into the eaves. Make sure you don't cover the ventilation gap (trim the end of the blanket to a wedge-shape so that it does not obstruct the airflow), or fit eaves vents.

Unroll the blanket between the joists, pressing it down to form a snug fit – but don't compress it. If you have bought a roll that's slightly wider than the joist spacing, allow it to curl up against the timbers on each side.

Continue at the opposite side of the loft with another roll. Cut it to butt up against the end of the first one, using either a large kitchen knife or a pair of long-bladed scissors. Continue across the loft till all the spaces are filled. Trim the insulation to fit odd spaces.

Do not cover the casings of light fittings that protrude into the loft space. Also, avoid covering electrical cables, as there is a risk that that could cause overheating. Instead, lay the cables on top of the blanket or clip them to the sides of the joists above it.

Do not insulate the area immediately below a cold-water cistern (the heat rising from the room below will help to prevent freezing during the winter).

Cut a piece of blanket to fit the cover of the entrance hatch, and attach it with PVA adhesive or with cloth tapes and drawing pins. Fit foam draught excluder around the edges of the hatch.

Laying loose-fill insulation

When laying loose-fill insulation, take precautions against condensation similar to those described for blanket insulation (see above). To avoid blocking the eaves, wedge strips of plywood or thick cardboard between the joists before laying the insulant.

Pour insulation between the joists and distribute it roughly with a broom. Level it with a spreader cut from hardboard. If the joists are shallow, nail on lengths of wood to build up their height to at least 150mm (6in), if only to support walkway boarding in specific areas of the loft. To insulate the entrance hatch, screw battens around the outer edge of the cover, then fill with granules and pin on a hardboard lid to contain them.

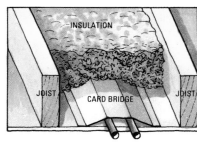

Insulating pipes between joists

INSULATING CISTERNS AND PIPES

Insulating cisterns
To comply with current bylaws, your cold-water-storage cistern must be insulated. It's simplest to buy a Bylaw 30 kit, which includes a cylinder jacket and all the other equipment that is required. Insulate your central-heating expansion tank at the same time.

Buy a ready-made jacket to insulate a cistern

Insulating pipes
If there are cold-water pipes running between the joists, lay the blanket insulation over them to prevent them from freezing. If that is not practical, insulate each pipe run separately.

Before pouring loose-fill insulation, lay a bridge made from thin card over cold-water pipes running between the joists, so they will benefit from warmth rising from the room below. If the joists are shallow, cover the pipes with foam sleeves before pouring the insulation.

Laying blanket insulation
(Left)
Seal all gaps around pipes, vents and wiring (**1**). Place end of roll against eaves and trim ends (**2**) or fit eaves vents (**3**). Press rolls between joists (**4**). Insulate cistern and cold-water pipes (**5**).

Spreading loose-fill insulant
(Right)
Seal gaps to prevent condensation (**1**). Use strips of plywood to prevent insulant from blocking ventilation (**2**) or fit eaves vents (**3**). Cover the cold-water pipes with a cardboard bridge (**4**), then use a spreader to level the insulant (**5**). Insulate and draughtproof the hatch cover (**6**).

Insulating between the rafters

If the attic to be used, you will need to insulate the sloping part of the roof in order to heat the living space. Repair the roof covering first, so the insulation won't be soaked by leaks (it will also be difficult to spot leaks after insulating).

Condensation often causes serious problems after installing insulation between the rafters, as the undersides of the roof tiles become very cold. It is therefore vital to provide a 50mm (2in) gap between the insulant and the tiles, to promote sufficient ventilation to keep the space dry (this also determines the maximum thickness of insulation you can install). The ridge and eaves must be ventilated, and you should include a vapour barrier on the warm side of the insulation – either by fitting foil-backed blanket or by stapling polyethylene sheet to the lower edges of the rafters to cover unbacked insulation.

After installing any type of insulation, you can cover the rafters with sheets of plasterboard as a final decorative layer. Your choice of panels will be limited by the maximum size you can pass through the hatchway of your loft. Fix the panels to the rafters with plasterboard nails or screws, staggering the joints.

An alternative is to provide insulation and surface finish together by fitting insulated (thermal) plasterboard to the underside of the rafters.

Fixing blanket insulant

Unfold the side flanges from a roll of foil-backed blanket and staple them to the underside of the rafters. As you fit adjacent rolls, make sure you overlap the edges of the vapour barrier in order to provide a continuous layer.

Attaching sheet insulant

The most satisfactory method is to cut sheet insulation accurately in order to ensure a wedge-fit between the rafters. If need be, screw battens to the sides of the rafters so you can fix the insulating sheets to them (treat the battens with preserver first). Install a polyethylene-sheet vapour barrier over the rafters, double-folding the joints before stapling them in place.

INSULATING AN ATTIC ROOM

If an attic room was built as part of the original dwelling, you will probably not be able to insulate the pitch of the roof unless you are prepared to hack off the old plaster before insulating between the rafters (see left). It may therefore be simpler to insulate from the inside (as for a flat roof), although you won't have a lot of headroom. Insulate the short vertical wall of the attic from inside the crawlspace, making sure the vapour barrier faces the warm inner side of the partition. At the same time, insulate between the joists of the crawlspace.

Fit blankets with vapour barrier facing the room

SEE ALSO

Insulating a room in the attic
Surround the room itself with insulation, but leave the floor uninsulated so that the attic will benefit from heat rising from the rooms directly below.

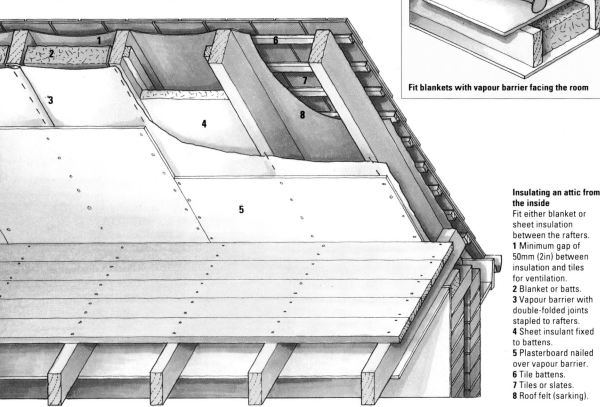

Insulating an attic from the inside
Fit either blanket or sheet insulation between the rafters.
1 Minimum gap of 50mm (2in) between insulation and tiles for ventilation.
2 Blanket or batts.
3 Vapour barrier with double-folded joints stapled to rafters.
4 Sheet insulant fixed to battens.
5 Plasterboard nailed over vapour barrier.
6 Tile battens.
7 Tiles or slates.
8 Roof felt (sarking).

INSULATING FLAT ROOFS AND WALLS

SEE ALSO
Details for:
External-wall insulation 77

Insulating a flat roof from outside
Expert contractors can insulate the roof from above.

Warm-roof system
1 Roof deck
2 Waterproof covering
3 New vapour barrier
4 Insulation
5 New waterproof covering

Protected-membrane system
1 Roof deck
2 Waterproof covering
3 Insulation
4 Paving slabs

Insulating the ceiling
Insulate a flat roof by fixing insulant to the ceiling.
1 Existing plasterboard or lath-and-plaster ceiling.
2 Softwood battens screwed to joists.
3 Insulation glued to existing ceiling.
4 Polyethylene vapour barrier stapled to battens.
5 Plasterboard nailed to battens.
6 If possible, provide cross ventilation by installing vents equal to 0.4 per cent of roof area.

Treatment from above

One way of insulating a flat roof is to lay rigid insulating board on the original deck. The bonded 'warm roof' system incorporates a vapour barrier – possibly just the old covering – that is laid under the insulation, which is then protected with a new waterproof covering. With a protected-membrane system, the insulation is laid over the covering and is held in place with paving slabs or a layer of pebbles. Both systems are best installed by contractors. Get them to check that the roof is weatherproof and can support the additional weight.

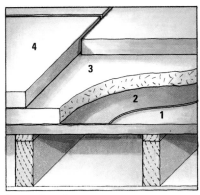

Warm-roof system

Protected-membrane system

Treatment from below

Another option is to insulate the ceiling below a flat roof. Very often the space within the roof structure has little or no ventilation. It is therefore essential to include a vapour barrier on the warm side of the ceiling, below the insulation, in order to prevent condensation.

Nail thermal plasterboard to the joists, or install fire-retardant expanded polystyrene, 50mm (2in) thick, between softwood battens screwed to the joists every 400mm (1ft 4in) across the ceiling. Fit the first batten against the wall at right-angles to the joists, then fit one at each end of the room. Butt the poly-styrene against the first batten; coat the back of the material with polystyrene adhesive, then glue it to the ceiling.

Continue with alternate battens and panels until you reach the other side of the room, finishing with a batten against the wall. Install a polyethylene vapour barrier, double-folding the joints and stapling them to convenient battens. Fix plasterboard panels to the battens with galvanized plasterboard nails. Stagger the joins between the panels, then fill the joins and finish ready for decorating as required. A double coat of oil paint is a vapour barrier itself.

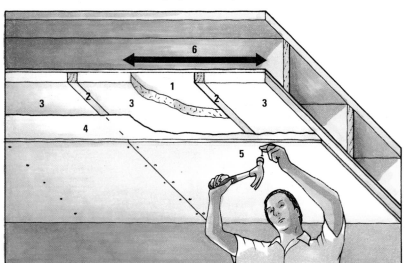

How you insulate the walls of your home is likely to be determined by several factors. Firstly, the type of construction. Houses built after 1920, and certainly after 1950, usually have cavity walls – two skins of brick, or one of brick and one of concrete block, with a gap between them through which air circulates to reduce the likelihood of water penetration. Although heat loss is slightly slower through a cavity wall than one of solid brick, that does not substantially reduce the cost of home heating. However, filling the cavity with insulation prevents circulation, trapping the air in millions of tiny air pockets within the material. This can reduce heat loss through the wall by as much as 65 per cent.

Solid walls require different treatment. You can either employ a contractor to insulate the external face of the walls or line the inner surfaces yourself.

Advantages and disadvantages

With cavity filling every exterior wall must be treated simultaneously, so it is most cost-effective for homes that are heated throughout for long periods and have a properly controlled system. Heating without controls will simply increase the temperature inside instead of saving on fuel bills. This type of insulation is not practical for flats or apartments unless the whole building is insulated at the same time.

Walls made of solid brick or stone have to be insulated in some other way. Cladding the exterior of the house with insulation is expensive and also ruins the appearance of most buildings. The suggestions concerning the manner of heating and effective controls to make cavity filling worthwhile apply equally to exterior-wall insulation.

Another method – suitable for solid and cavity walls – is to line the inner surfaces of the walls with insulation. This may involve a great deal of effort, depending on the amount of alteration required to joinery, electrical fittings and plumbing – but it does provide an opportunity for selective insulation, concentrating on those rooms which are likely to benefit most. It is also the only form of wall insulation that can be carried out by the householder.

INSULATING
WALLS

SEE ALSO
Details for:
Furring strips   77

When constructing a new house, a builder will include a layer of insulation between the two masonry leaves of exterior walls – a simple measure that greatly increases the thermal insulation of the building. To insulate an existing wall is a different matter. It requires a skilled and experienced contractor to introduce an insulant through holes cut in the outer brick leaf and to fill the cavity in such a way that a substantial reduction of heat loss is achieved, while avoiding the undesired side effect of damp penetrating to the inner leaf.

It's advisable to hire contractors that are approved by the Agrément Board or registered with the British Standards Institution, or belong to the National Cavity Insulation Association.

You should expect the contractor to carry out a thorough initial survey of the building to make sure that the walls are structurally fit for filling and that there is no evidence of frost damage or failed pointing. An approved company will also make the necessary application to the local authority before commencing installation, in order to comply with the Building Regulations. This is particularly important if you live in an area of the country where your house is exposed to severe driving rain or blizzards for prolonged periods, since not all cavity fillings are suitable for such extreme weather conditions.

Do not hire a contractor who does not provide a long-term guarantee that is transferable with ownership of the house. It should state that the insulant will be effective throughout the period of the guarantee and that it is rot-proof

and vermin-proof. Most important of all perhaps, check that the contractor's guarantee unequivocally states that damp resulting from faulty material or installation will be cured free of charge.

You are most likely to be offered one of three insulants. Urea-formaldehyde foam is the cheapest and most widely used material, despite its reputation for releasing unpleasant odours as the foam cures. In fact this happens in very few instances, and it is then usually due to a poor survey that has failed to detect that the walls of the house are not in a fit condition for filling.

Mineral or glass fibre treated with a water repellent are the next most popular cavity insulants. Both materials are completely inert and, if properly installed, will form a stable insulation that will neither settle nor shrink once inserted into the cavity.

Expanded-polystyrene beads form the third most commonly used insulant. Some of them are lightly coated with adhesive at the moment of injection, so that the fill does not settle over a period of time. If polystyrene is treated and properly installed, it does not affect the fire resistance of a masonry wall.

Whatever process you choose, the work should take no more than two to three days to complete and all of it is carried on outside the house, where holes are drilled at regular intervals in the brickwork (1). The insulant is either injected or blown through a hose (2), then the insertion holes are plugged. If the work is done properly, the holes should be virtually invisible, except perhaps on close inspection.

DRY LINING FROM THE INSIDE

If you're planning to dry line an external wall with some form of panelling, it is worth taking the opportunity to include blanket or sheet insulation between either the wall battens or the furring strips. Fix a polyethylene-sheet vapour barrier over the insulation by stapling it to the furring strips before nailing the panelling in place. Alternatively, use a metallized plastic-backed plasterboard, which has the advantage of requiring no additional vapour barrier.

Any form of panelling can be applied over mineral-fibre or glass-fibre blanket insulation, but plasterboard should be used to cover expanded-polystyrene insulant. A somewhat simpler method is to glue insulated (thermal) plasterboard directly onto a sound plaster surface. This type of wall insulation is made from standard plasterboard backed by either a layer of expanded-polystyrene or rigid-polyurethane foam. An integral vapour barrier is incorporated in both boards (see below).

Using a notched applicator, apply a band of the manufacturer's adhesive, 200mm (8in) wide, to the wall so as to coincide with the vertical edges of the panel and its centre line (1). Spread horizontal bands top and bottom. Press the panel against the adhesive and tamp it down with a heavy straightedge, then secure it to the wall with nine nailable plugs (2) in three rows. Position the plugs 50mm (2in) from all edges of the panel, and place one in the centre. Use a fine-toothed saw to cut a panel to fit into a corner. At a cut edge, allow for a 3mm (⅛in) gap for filling after the panel is fixed. Tape and fill all joints.

Details regarding door and window mouldings can be found in the section on wall panelling, but bed the skirting board onto a bead of mastic sealant applied to the floor and plasterboard. Fix the skirting board through the panel to the wall behind.

1 Fixing thermal plasterboard
Spread adhesive along the bands shown in the diagram above. The crosses indicate the centres of nailable fixing plugs.

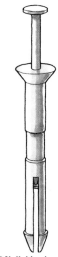

2 Nailable plug
Used to fix insulated plasterboard to the wall. Push the plug into a hole drilled through the board, then drive in the nail to expand the plug and grip the masonry.

1 Drilling holes in the outer leaf
A professional contractor will begin by drilling large-diameter holes through the outer skin of brickwork to gain access to the cavity.

2 Introducing the insulant
A hose is then inserted into each hole and the insulant is injected or blown into the cavity under pressure, filling it from the base. Afterwards, the holes are plugged with colour-matched mortar.

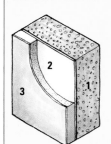

1 Insulant (either expanded polystyrene or foamed polyurethane)
2 Integral vapour barrier
3 Plasterboard lining

The structure of insulated plasterboard

LAGGING PIPES, CYLINDERS AND RADIATORS

SEE ALSO
Details for:
Elbow fitting   77

Insulating a hot-water cylinder

Many people think that an unlagged cylinder has the advantage of providing a useful source of heat in an airing cupboard – but in fact it squanders a surprising amount of energy. Even a lagged cylinder should provide ample heat in an enclosed airing cupboard; if not, an uninsulated pipe will do so.

Proprietary water-cylinder jackets are made from segments of mineral-fibre insulation, 80 to 100mm (3¼ to 4in) thick, wrapped in plastic. Measure the approximate height and circumference of the cylinder to choose the right size.

If need be, buy a jacket that is too large, rather than one that is too small. Make sure the quality is adequate by checking that it is marked with the British Standard kite mark (BS 5615).

Thread the tapered ends of the jacket segments onto a length of string and tie it round the pipe at the top of the cylinder. Distribute the segments evenly around the cylinder and wrap the straps or tapes provided around it to hold the jacket in place.

Spread out the segments to make sure the edges are butted together, and tuck the insulation around the pipes and the cylinder thermostat.

If you should ever have to replace the cylinder itself, consider substituting a pre-insulated version, of which there are various types on the market.

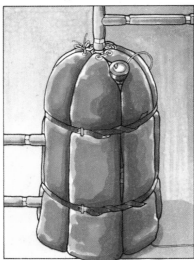

Lagging a hot-water cylinder
Fit a jacket snugly around the cylinder and wrap foamed-plastic tubes (see above right) around the pipework, especially the vent pipe directly above the cylinder.

Lagging pipe runs

You should insulate hot-water pipes in those parts of the house where their radiant heat is not contributing to the warmth of the rooms, and cold-water pipes in unheated areas of the building (where they could freeze). You can wrap pipework in lagging bandages (there are several types, some of which are self-adhesive), but it is generally more convenient to use foamed-plastic tubes designed for the purpose. This is especially true for pipes close to a wall, which may be awkward to wrap.

Foamed-plastic tubes are produced to fit pipes of different diameters: the tube walls vary in thickness from 12 to 20mm (½ to ¾in). More expensive varieties incorporate a metallic-foil backing that reflects some of the heat back into hot-water pipes.

Most tubes are pre-slit along their length so that they can be sprung over the pipe (1). Butt successive lengths of tube end-to-end, and seal the joints with PVC adhesive tape.

At a bend, cut small segments out of the split edge so that it bends without crimping. Fit it around the pipe (2) and seal the closed joints with tape. If two pipes are joined with an elbow fitting, mitre the ends of the two lengths of tube, butt them together (3) and seal with tape. Cut lengths of tube to fit snugly around a tee-joint, linking them with a wedge-shaped butt joint (4), and seal with tape as before.

| 1 Spring onto pipe | 2 Cut to fit bend | 3 Mitre over elbows | 4 Butt at tee-joint |

Reflecting heat from a radiator

As much as 25 per cent of the radiant heat from a radiator against an outside wall may be lost to the wall behind it. You can reclaim maybe half this wasted heat by applying a foil-faced expanded-polystyrene lining to the wall behind the radiator, to reflect the heat back into the room. The material is available as rolls, sheets or tiles; and although it is easiest to apply it to the wall when the radiator is removed for decorating, that is certainly not essential.

Turn off the radiator and measure it, including the position of the brackets.

Use a sharp trimming knife or scissors to cut the lining to size, so it is slightly smaller than the radiator all round. Cut narrow slots, as need be, to fit over the fixing brackets (1).

Apply heavy-duty fungicidal wall-paper paste to the back of the material and then slide it behind the radiator (2). Smooth it onto the wall with a radiator roller or wooden batten. Allow the paste to dry before turning the radiator on again. Alternatively, you can fix the lining in place with double-sided adhesive pads.

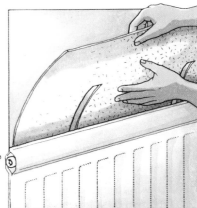

| 1 Cut slots to align with wall brackets | 2 Slide lining behind radiator and press to wall |

A double-glazed window consists of two sheets of glass separated by an air gap. The air gap provides an insulating layer, which reduces heat loss and sound transmission. Condensation is also reduced because the inner layer of glass remains warmer than the glass on the outside. Factory-sealed units and secondary glazing are the two methods commonly used for domestic double glazing. Both will provide good thermal insulation. Sealed units are unobtrusive, while secondary glazing offers improved noise insulation. Choose the type that suits your requirements best.

What size air gap?

For heat insulation, a 20mm (¾in) gap will give the optimum level of efficiency. If the gap is less than 12mm (½in), the air can conduct a proportion of the heat across it. If it's greater than 20mm (¾in), there is no appreciable gain in thermal insulation, and air currents can transmit heat to the outside layer of glass. For noise insulation, an air gap of 100 to 200mm (4 to 8in) is more effective. A combination of a sealed unit coupled with secondary glazing, known as triple glazing, provides the ideal solution.

Double glazing will help to cut your fuel bills, but the benefit that you will be aware of immediately is the elimination of draughts. The cold spots associated with a large window, which are most noticeable when you are sitting still, will also be reduced.

In terms of saving energy, the heat lost through windows is relatively small – around 10 to 12 per cent – compared to the heat loss for the whole house, but the installation of double glazing can halve this amount.

Double glazing will improve security against forced entry, particularly when sealed units or toughened glass have been installed. However, make sure that some accessible part of the windows can be opened to afford an emergency escape route in case of fire.

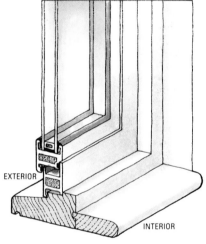

Factory-sealed unit
A complete frame system installed by a contractor.

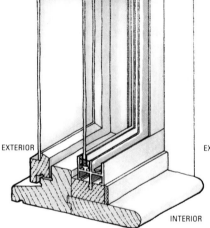

Secondary double glazing
Fitted in addition to an ordinary glazed window.

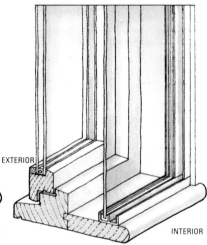

Triple glazing
A combination of secondary and sealed units.

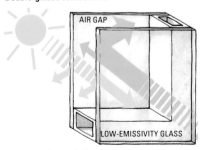

Double-glazed sealed units

Heat-retentive sealed unit

Double-glazed sealed units

Double-glazed sealed units consist of two panes of glass that are separated by a spacer and hermetically sealed all round. The cavity between the glass may be 6, 9, 12 or 20mm (¼, ⅜, ½ or ¾in) wide. The gap may contain dehydrated air – which eliminates condensation between the two panes of glass – or inert gases, which also improve thermal and acoustic insulation.

The thickness and type of glass used are determined by the size of the unit. Clear float glass or toughened glass is commonly employed. When obscured glazing is required to provide privacy, patterned glass is used. Heat-retentive sealed units, incorporating special low-emissivity glass, are supplied by some double-glazing companies,

Generally, factory-sealed units are produced and installed by suppliers of ready-made double-glazed replacement windows. You can also buy sealed units that are suitable for self-fixing from some joinery suppliers, or have them made to order by specialists. Square-edged units are available for frames with a deep rebate, and stepped units for window frames that were originally intended for single glazing.

Double-glazed sealed units with PVC or aluminium frames are rarely suitable for older houses. A secondary system that leaves the original window intact is generally more appropriate – especially if you have attractive leaded windows, which need to be preserved (a sealed unit with fake glazing bars or a modern interpretation of leaded lights is not an adequate substitute for the real thing).

65

SECONDARY DOUBLE GLAZING

● **Providing a fire escape**
If you fit secondary glazing, make sure there is at least one window in every occupied room that can be opened easily.

Secondary double glazing comprises a separate pane of glass or plastic sheet fitted over an ordinary single-glazed window. It is normally fitted on the inside of the existing window, and is one of the most popular methods of double glazing as it is relatively easy to install yourself – usually at a fraction of the cost of other systems.

How the glazing is fixed

Secondary glazing can be fastened to the sash frames (1) or window frame (2), or across the window reveal (3). The method depends on the ease of fixing, the type of glazing chosen, and amount of ventilation required.

Glazing fixed to the sash will reduce heat loss through the glass and provide accessible ventilation, but it won't stop draughts – whereas glazing fixed to the window frame has the advantage of cutting down heat loss and eliminating draughts at the same time. Glazing fixed across the reveal offers improved noise insulation too, since the air gap can be wider. Any system should be readily demountable, or preferably openable, to provide a change of air if the room does not have any other form of ventilation.

A rigid-plastic or glass pane can be fitted to the exterior of the window if secondary glazing fitted on the inside would look unsightly. Windows set in a deep reveal, such as the sliding-sash type, are generally the most suitable ones for external secondary glazing (4).

Glazing with renewable film

Quite effective double glazing can be achieved using double-sided adhesive tape to stretch a thin flexible sheet of plastic across a window frame. The taped sheet can be removed at the end of the winter.

Clean the window frame (1) and cut the plastic roughly to size, allowing an overlap all round. Apply double-sided tape to the edges of the frame (2), then peel off the backing paper.

Attach the plastic film to the top rail (3), then tension it onto the tape on the sides and bottom of the window frame (4). Apply only light pressure until you have positioned the film, and then rub it down onto the tape all round.

Remove all creases and wrinkles in the film, using a hair dryer set to a high temperature (5). Starting at an upper corner, move the dryer slowly across the film, holding it about 6mm (¼in) from the surface. When the film is taut, cut off the excess plastic (6).

GLAZING POSITIONS

Secondary double glazing is particularly suitable for DIY installation. It's possible to fit a secondary system to almost any style or shape of window.

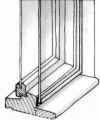

1 Sash-fixed
Glazing fixed to the opening window frame.

2 Frame-fixed
Glazing fixed to the structural frame.

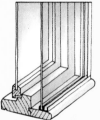
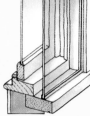

3 Reveal-fixed
Glazing fixed to the reveal and interior windowsill.

4 Exterior-fitted
Glazing fixed to the reveal and exterior windowsill.

1 Wipe woodwork to remove dust and grease

2 Apply double-sided tape to the fixed frame

3 Stretch the film across the top of the frame

4 Pull the film tight and fix to sides and bottom

5 Use a hair dryer to shrink the film

6 Trim the waste with a sharp knife

Demountable systems

A simple method of interior secondary glazing uses clear-plastic film or sheet. These lightweight materials are held in place by self-adhesive strips or rigid moulded sections, which form a seal. Most strip fastenings use magnetism or some form of retentive tape, thus allowing the secondary glazing to be removed for cleaning or ventilation. The strips and tapes usually have a flexible-foam backing, which takes up slight irregularities in the woodwork. This type of glazing can be left in place throughout the winter and removed for storage during the summer months.

Fitting a demountable system

Clean the windows and the surfaces of the window frame. Cut the plastic sheet to size. Place the glazing on the window frame and mark around it **(1)**. Working with the sheet on a flat table, peel back the protective paper from one end of the self-adhesive strip. Apply the strip to the surface of the plastic, flush with one edge. Cut it to length and repeat on the other edges. Cut the mating parts of the strips and apply them to the window frame following the guide lines. Press the glazing into place **(2)**.

When using rigid moulded sections, cut the sections to length with mitred corners. To fit an extruded clip-type moulding **(3)**, stick the base section to the frame, then insert the outer section to retain the glazing.

1 Mark around glazing   **2 Position glazed unit**

SELF-ADHESIVE
TAPE

GLAZING

PLASTIC MOULDINGS

3 Rigid plastic mouldings support the glazing

PLASTIC MATERIALS FOR DOUBLE GLAZING

To reduce costs or for safety, plastic materials can be used in place of glass to provide lightweight double glazing. They are available as clear, thin flexible film and as clear, textured or coloured rigid sheets.

Unlike glass windows, plastic glazing has a high impact-resistance and does not splinter when broken. Depending on its thickness, plastic can be cut with scissors, drilled, sawn, planed or filed.

The clarity of new plastics is as good as glass, but they are easily scratched. They are also liable to degrade with age and are prone to static. To clean plastic sheet, wash with a liquid-soap solution. Slight abrasions can be rubbed out with metal polish.

Film and semi-rigid plastics are sold by the metre or in rolls. Rigid sheets are available in a range of standard sizes or can be cut to order.

Rigid-plastic sheets are generally supplied with a protective covering of paper or thin plastic on both faces. To avoid scratching the surface of the sheets, only peel off the covering after cutting and shaping.

Polyester film

Polyester film is a form of plastic often used for inexpensive secondary double glazing. It can be trimmed with scissors or a knife, and fixed with self-adhesive tape or strip fasteners. Since polyester is tough, virtually tearproof and very clear, it is an ideal plastic for glazing living-room windows. It is sold in 5, 10 and 25m (32, 64 and 160ft) rolls, 1143mm (3ft 9in) and 1270mm (4ft 2in) wide.

Polystyrene

Polystyrene is an inexpensive clear or textured rigid plastic. Clear polystyrene doesn't have the clarity of glass and degrades in strong sunlight – so should not be used for south-facing windows or if a distortion-free view is desirable. Depending on the climate, the life of polystyrene is estimated to be between three and five years. Its working life can be extended if the glazing is removed for storage in summer. It is available in thicknesses of 2mm ($\frac{1}{16}$in), 3mm ($\frac{1}{8}$in) and 4mm ($\frac{5}{32}$in), and in sheet sizes up to 1372 x 2440mm (4ft 6in x 8ft).

Acrylic

Acrylic is a good-quality rigid plastic. It is up to ten times stronger than glass, but without any loss in clarity.

Although acrylic costs about twice as much as polystyrene, its working life is estimated to be at least 15 years and it is manufactured in a useful range of translucent and opaque colours.

The thicknesses commonly available for clear glazing are 2.5mm ($\frac{3}{32}$in) and 3mm ($\frac{1}{8}$in). It is produced in sheet sizes up to 1220 x 2440mm (4ft x 8ft).

Polycarbonate

A lightweight vandal-proof glazing with a high level of clarity, polycarbonate is most commonly available as twin-wall sheeting for glazing conservatory roofs, although a triple-wall version is also produced. The hollow ribbed section gives it exceptional rigidity, at the same time keeping heat loss and weight to a minimum. Polycarbonate is available in 4, 6, 8,10 and16mm ($\frac{1}{8}$, $\frac{1}{4}$, $\frac{5}{16}$, $\frac{3}{8}$ and $\frac{5}{8}$in) thicknesses, and is sold in sheet sizes up to 4000 x 1250mm (13 x 4ft 1in).

PVC glazing

PVC is available as a flexible film or rigid sheet that is ultraviolet-stabilized and therefore unaffected by sunlight. PVC film provides inexpensive glazing where a high degree of clarity is not essential (for example, in a bedroom).

Rigid PVC, which is 3mm ($\frac{1}{8}$in) thick, is produced in sheet sizes up to 1220mm x 2440mm (4ft x 8ft). It can be used to glaze conservatories and carport roofs.

Details for:
Secondary
double glazing   66, 68

OPENABLE
SECONDARY
GLAZING

Hinged system
1 Glazing
2 Glazing gasket
3 Corner joints
4 Hinges
5 Aluminium extrusion
6 Turn button
7 Draughtproofing strip

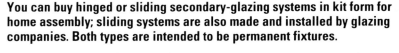

You can buy hinged or sliding secondary-glazing systems in kit form for home assembly; sliding systems are also made and installed by glazing companies. Both types are intended to be permanent fixtures.

Types of glass used

Normally 4mm ($\frac{5}{32}$in) glass is used in an openable secondary-glazing system.

For sliding windows, no pane should exceed 1.85sq m (20sq ft). In side-hung hinged sections the panes should be no more than 1.1sq m (12sq ft), but panes that are top-hung can be 1.65sq m (18sq ft). The height of each pane should not exceed 1.5m (5ft), nor should the height be more than twice the width.

For low windows or those that are at risk from impact, use toughened glass.

Hinged system

Hinged systems incorporate aluminium extrusions to form a frame for the glass or rigid-plastic sheet. The glazing sits in a flexible gasket lining the extrusions. Screw-fixed corner joints hold the sides of the frame together, and pivot hinges are inserted into one of the extrusions to make side-hung or top-hung units. Hinged units are fitted to the face of a wooden window frame and secured by turn buttons. A flexible draughtproofing strip is fixed to the back of the frame. A self-locking stay can be fitted to keep the window open to provide ventilation.

Sliding systems

A horizontally sliding glazing system is normally used for casement windows, whereas a vertically sliding system is more suitable for tall windows such as double-hung sashes. Both rigid-plastic and aluminium versions are available. Each of the panes is framed by a light-weight extrusion, which is jointed at the corners, and the glass is sealed into its frame with a gasket.

A horizontal system has two or more sliding panes, the number depending on the width of the window. They are held in a tracked frame, which is screwed to the window frame or the reveal. Fibre seals are fitted to the sliding-frame members to prevent draughts between the moving parts. The glazing is opened with an integral handle, and each pane can be lifted out for cleaning.

A vertically sliding system has much the same form of construction, but the frame incorporates ratchet catches to hold the panes open at any height.

Fixing a horizontally sliding system

Measure your window opening, and buy a kit of parts slightly larger than the opening. After cutting the vertical track members to size using a junior hacksaw (**1**), plug and screw them to either the reveal or the inside face of the window frame. Then cut the horizontal track members and screw them in place (**2**).

Measure the opening for the glazing and have it cut to size, following the manufacturer's instructions regarding tolerances. Arrange the system so that the overlapping members of the sliding frames coincide with vertical window mullions. Cut and fit the components of the glazing frame, including the gaskets and seals. Join the four sides together, usually with screw-fixed corner joints (**3**), and lift the glazing into the sliding tracks to complete the installation (**4**).

Sliding system
1 Glazing extrusion
2 Glazing
3 Corner joints
4 Glazing gasket
5 Top track
6 Bottom track
7 Side track
8 Fibre draught seal
9 Slides

1 Cut track to length	2 Screw it to the frame	3 Assemble the frame	4 Fit it into the tracks

Ventilation is essential for a comfortable atmosphere, but it has a more important function with regard to the structure of our homes. It wasn't a problem when houses were heated with open fires, drawing fresh air through all the natural openings in the structure; but with central heating and thorough insulation and draughtproofing, well-designed ventilation is vital. Without a constant change of air, centrally heated rooms become stuffy, and the moisture content of the air soon becomes so high that water is deposited as condensation – often with serious consequences. There are various ways to provide ventilation – some extremely simple, others much more sophisticated, giving total control.

Initial considerations

Whenever you plan an improvement to your home that involves insulation in one form or another, take into account how it's likely to affect your existing ventilation. It may change conditions sufficiently to create a problem in areas outside the habitable rooms, so that damp and its side effects are able to develop unnoticed under floorboards or in the loft. If there is any likelihood that damp conditions might occur, provide additional ventilation.

Fitting a fixed window vent

You can provide continuous ventilation by installing an inexpensive fixed vent in a window. Well-designed ventilators of this kind allow a free flow of air without causing draughts (they usually have a wind shield on the outside), and are totally reliable as there are no moving parts to break down or produce those irritating squeaks that are associated with wind-driven fans.

Have a glazier cut the recommended size of hole in the glass. Then fit one of the vent's louvred grilles on each side of the window (clamping them together with the central fixing bolt), and bolt the plastic windshield to the outer grille.

Ventilating a fireplace

An open fire needs oxygen to burn well. If the air supply is reduced by thorough draughtproofing or double glazing, then the fire smoulders and the slightest downdraught will blow smoke into the room. There may be other reasons why a fire burns poorly – such as a blocked chimney flue – but if you find that the fire picks up within a few minutes of partially opening the door to the room, you can be certain that inadequate ventilation is the cause of the problem.

The most efficient solution is to cut a hole in the floorboards on each side of the hearth and cover the hole with a ventilator grille. Although cheap plastic grilles are just as effective, brass or aluminium ones look more attractive in a living room. Choose a hit-and-miss ventilator, which you can close to seal off draughts when the fire is not in use. If you have a fitted carpet, cut a hole in it and screw the grille on top.

If the room has a solid floor, a simple alternative is to fit a grille over the door or above a window. An aperture at that height won't create a draught, because cold air will disperse across the room and will warm up as it falls.

Ventilating an unused fireplace
A fireplace that has been blocked with brickwork, blockwork or plasterboard should be ventilated so air can flow up the chimney to dry out condensation or penetrating damp. Some people believe a vent from a warm interior aggravates the problem by introducing moist air to condense on the cold surface of the flue. However, so long as the chimney is uncapped, the updraught should draw moisture-laden air to the outside.

An airbrick cut into the flue from outside is a safer solution, although it is much more difficult to accomplish – and quite impossible if you live in a terraced house. Moreover, the airbrick will have to be either blocked or replaced should you later decide to reopen the fireplace.

To ventilate the fireplace from inside the room, either remove a single brick, make an aperture in the blocks, or cut a hole in the plasterboard used to block off the fireplace. Screw a face-mounted ventilator over the hole, or use one that is designed for plastering in (the thin flange for screw-fixing the ventilator to the wall will be covered as you plaster up to the slightly protruding grille).

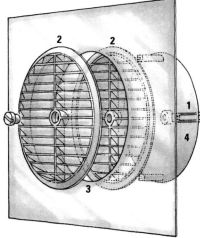

The components of a fixed window vent
1 Fixing bolt
2 Louvred grilles
3 Hole in the glass
4 Wind shield

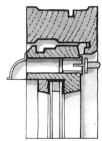

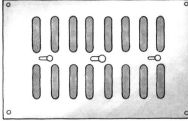

Hit-and-miss ventilator grille

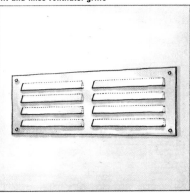

Face-mounted grille for ventilating a fireplace

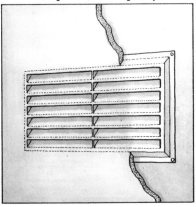

Hide the fixings of a grille with plaster

Trickle ventilation
Replacement windows can be supplied with a slot (either at the top of the fixed frame or in the movable sash) for a controllable trickle ventilator. This type of ventilator can also be fitted to an existing window in order to provide background ventilation when the window is closed.

69

VENTILATING BELOW FLOORS

Perforated openings known as airbricks are built into the external walls of a house to ventilate the space below suspended wooden floors. If they become clogged with earth or leaves, there's a strong possibility of dry rot developing in the timbers, so check their condition regularly. Clear a blocked airbrick as soon as you discover it – and if the original ones are inadequate, replace them with larger new ones.

Checking out the airbricks

Ideally there ought to be an airbrick every 2m (6ft) along an external wall, but in a great many buildings there is less provision for ventilation without ill effect – in fact sufficient airflow is more important than the actual number of openings in the wall.

Floor joists that span a wide room are supported at intervals by low sleeper walls of brick. Sometimes these are perforated to facilitate an even air-flow throughout the space – but in other cases there are merely gaps left by the builder between sections of solid wall. This method of constructing sleeper walls can lead to pockets of still air in corners where draughts never reach.

Even when all the airbricks are clear, dry rot can break out in areas that don't receive an adequate change of air. If you suspect that there are 'dead' areas under your floor – particularly if there are signs of damp or mould growth – fit an additional airbrick in a wall nearby.

Old ceramic airbricks do get broken, and are often ignored because there is no detrimental effect on the ventilation. However, even a small hole provides access for vermin. Don't be tempted to block the opening, even temporarily, but replace the broken airbrick with a similar one of the same size. You can choose from single or double-size airbricks, made in ceramic or plastic.

Installing or replacing an airbrick

Use a masonry drill to excise the mortar surrounding the brick you are removing, and a cold chisel to chop out the brick itself. You may have to cut some bricks to install a double-size vent. Having cut through the wall, spread mortar on the base of the hole and along the top and both sides of the new airbrick. Push it into the opening, keeping it flush with the face of the brickwork, then repoint the mortar to match the profile used on the surrounding wall.

Single ceramic brick

Double-size plastic airbrick

Ventilating the space below a suspended wooden floor
The illustration (right) shows a cross-section through a typical cavity-wall structure with a wooden floor suspended over a concrete base. A house with solid-brick walls is ventilated in a similar way.
1 Airbrick fitted with telescopic sleeve.
2 Sleeper wall built with staggered bricks in order to allow air to circulate.
3 Floorboards and joists are susceptible to dry rot caused by poor ventilation.

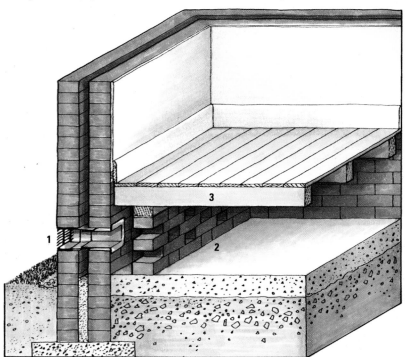

BRIDGING A CAVITY WALL

To build an airbrick into a cavity wall, bridge the gap with a plastic telescopic unit, which is mortared into the hole from both sides. If need be, a ventilator grille can be screwed to the inner end of the telescopic unit.

Where an airbrick is inserted above the DPC, you must fit a cavity tray over the telescopic unit to prevent water from percolating to the inner leaf of the cavity wall.

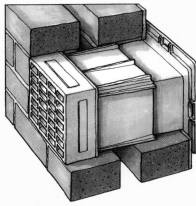

Airbrick with telescopic sleeve
Bridge a cavity wall with this type of unit.

Cavity tray
A cavity tray sheds any moisture that penetrates the cavity above the unit. It is necessary only when the airbrick is fitted above the DPC.

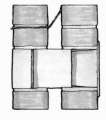

VENTILATING AN APPLIANCE

A fuel-burning appliance with a flue must have an adequate supply of air to function efficiently and safely. If any alteration or improvement interferes with that supply, then you must provide alternative ventilation. If you plan to block a vent, or alter a window or install an extractor fan in the same room as the appliance, consult a professional fitter. He will tell you whether the alteration is advisable, what type and size of vent to install, and where it should be positioned for best effect. An appliance with a balanced flue draws air directly from outside the house, so will not be affected by internal alterations.

When loft insulation first became popular as an energy-saving measure, householders were recommended to tuck insulant right into the eaves to keep out draughts. What people failed to recognize was that a free flow of air is necessary in the roof space to prevent moisture-laden air from below condensing on the structure. Inadequate ventilation can lead to serious deterioration: wet rot develops in the roof timbers and water drops onto the insulant, eventually rendering it ineffective as insulation. If water builds up into pools, the ceiling below becomes stained and there is a risk of short-circuiting the electrical wiring in the loft. For these reasons, efficient ventilation of the roof space is essential in every home.

Ventilating the eaves

The regulations applicable to new housing insist on ventilation equivalent to continuous openings of 10mm (⅜in) along two opposite sides of a roof with a pitch (slope) of 15 degrees or more. If the pitch of the roof is less than 15 degrees, ventilation should amount to the equivalent of 25mm (1in) continuous openings. It also makes sense to adopt similar standards when refurbishing a house of any age.

The simplest method of ventilating a standard pitched roof is to fit round soffit vents made with integral insect screens. The spacing is determined by the size of opening provided by the particular vent. Push the vents into openings cut with a hole saw.

If the opening at the eaves is likely to be restricted by insulation, insert a plastic or cardboard eaves vent between each pair of joists. Push the vent into the angle between the rafters and the joists, with the ribbed section uppermost. Vents can be cut to length with scissors for an exact fit. When installing blanket or loose-fill insulation, push it up against the vent.

Slate and tile vents

Certain types of roof construction do not lend themselves to ventilation from the eaves only, but the structure can be ventilated successfully by replacing strategic tiles or slates with specially designed roof vents. A range of colours and shapes is available to blend with various roof coverings.

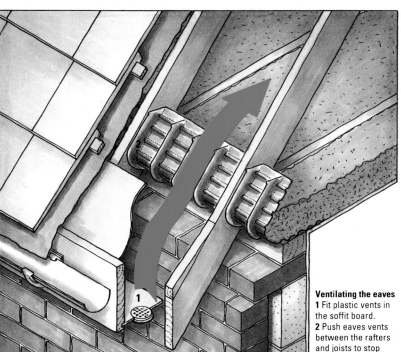

Ventilating the eaves
1 Fit plastic vents in the soffit board.
2 Push eaves vents between the rafters and joists to stop insulation blocking the flow of air.

WHEN ROOF VENTS ARE ESSENTIAL

Eaves-to-eaves ventilation normally keeps the roof space dry, but it is sometimes necessary to fit tile or slate vents to draw air through the roof space.

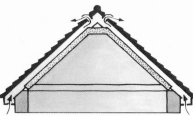

An attic space
If you insulate the slope of your roof, you must provide a minimum 50mm (2in) airway between the insulant and roof-covering. Fit soffit vents at the eaves, plus slate or tile vents near the ridge of the roof.

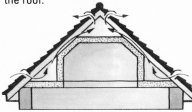

A room in the roof
Where a room is built into the attic, fit vents near the eaves and ridge to draw air through the narrow spaces over the sloping ceiling.

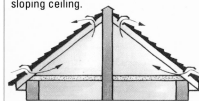

A fire or party wall
A solid wall built across the loft space prevents eaves-to-eaves ventilation. Fit slate or tile vents to ventilate each side of the wall independently. Use the same arrangement to ventilate a mono-pitch roof over an extension or lean-to.

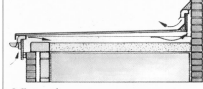

A flat roof
An insulated flat roof can be ventilated by fitting over-fascia ventilators at the eaves and at the wall abutment. Some modification of the wall flashing will be necessary on an existing roof.

SEE ALSO

Details for:	
Wet rot	15
Insulation	60
Insulating an attic	61
Insulating sloping roof	61
Fitting roof vents	72

Soffit vent

Slate/tile vents
Roof vents are made to resemble a variety of roof coverings.

Double Roman tile vent

Slate vent

Double pantile vent

Over-fascia ventilator

FITTING ROOF VENTS

Installing a slate or tile vent

If you are having the roof of your house replaced, it is worth getting the roofing contractor to incorporate vents at the same time – but it is possible to install roof vents yourself. Fitting instructions for individual models will vary in detail, but the description below for a double pantile vent demonstrates the principle.

As each vent must be fitted between rafters, you need to select the approximate position for a vent and remove just enough tiles to expose the felt and locate the heads of the nails holding the tile-support battens to the roof. The nails indicate the position of the rafters.

Centre the template supplied by the manufacturer between the rafters, and mark the position of the hole on the roof felt by scratching the corners with a knife (1). Cut the diagonals of the opening and bend back the flaps (2). Cut a slit in the felt, 100mm (4in) above and centred on the opening, and insert the tail of the undercloak of the vent (3); then align the edge of the undercloak with the opening (4). Plug the extension sleeve onto the underside of the cowl (5), then insert the sleeve into the hole in the felt and nail the cowl to the support batten above the opening (6). Replace the surrounding tiles, using hook clips to hold them in position.

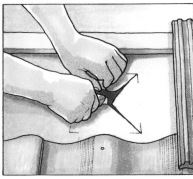

1 Scratch the corners of the hole with a knife

2 Slice the felt diagonally to make four flaps

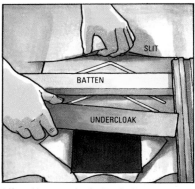

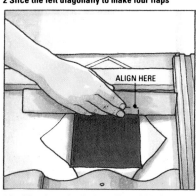

3 If need be, put undercloak under batten

SLIT
BATTEN
UNDERCLOAK

4 Position the undercloak to align with hole

ALIGN HERE

The undercloak
The undercloak fits beneath the vent cowl in order to prevent water running through the opening cut in the roof felt.

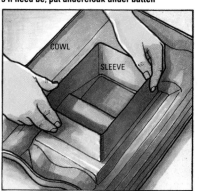

COWL
SLEEVE

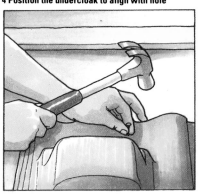

5 Plug sleeve onto the bottom of the vent cowl

6 Nail the vent to the tile batten

CALCULATING AIR-FLOW CAPACITY

All vents positioned near the eaves should provide the equivalent of a 10 or 25mm (⅜ or 1in) continuous gap, the size depending on the pitch of the roof. The ones near the ridge should provide airflow to suit the roof's construction.

To calculate how many vents will be needed, divide the specified airflow capacity of the type of vent you wish to use into the recommended continuous gap. If you are in doubt, provide slightly more ventilation than is indicated.

Position eaves vents in the fourth or fifth course of slates or tiles; place the higher vents a couple of courses below the ridge. Space all vents evenly along the roof, to ensure that there aren't any areas of 'dead' air.

CLEARING THE OPENING

When replacing tiles or slates with a vent you may have to cut through a tile-support batten to clear the opening. Nail a short length of batten above and below the opening to provide additional support. Because roofing slates overlap each other by a considerable amount, you'll find that you have to cut away the top corners of the lapped lower slates.

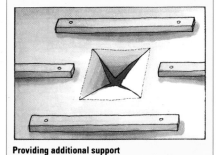

Providing additional support
Cut a tile batten that obstructs a hole, then place battens above and below to support the vent.

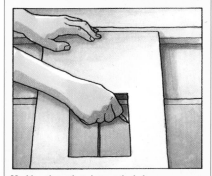

Marking slates that obstruct the hole
When slates cover the opening for a vent, use the template to mark the corners, then remove them.

FITTING AN EXTRACTOR FAN

SEE ALSO

Details for:	
Condensation	10-11
Cooker hoods	75

Since kitchens and bathrooms are particularly prone to condensation, it is important to have some means of expelling moisture-laden air together with unpleasant odours. An electrically driven extractor fan freshens a room faster than if you have to rely on natural ventilation and without creating draughts. This type of fan is now mandatory in new kitchens and bathrooms.

Positioning an extractor fan

The best place to site a fan is either in a window or on an outside wall, but its exact position is more critical than that. Stale air extracted from the room must be replaced by fresh air – normally through the door leading to other areas of the house. But if the fan is sited close to the source of replacement air, it will promote local circulation while having little effect on the rest of the room. The ideal position for it is directly opposite the source of replacement air, as high as possible, to extract the hot air (1). In a kitchen, try to locate the fan adjacent to the cooker, so that cooking smells and steam will not be drawn across the room before being expelled (2).

If the room contains a fuel-burning appliance with a flue, you must ensure there is an adequate supply of fresh air at all times. The only exception is an appliance with a balanced flue, which takes its air directly from outside.

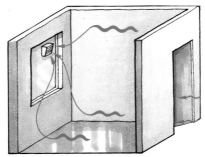

1 Fit extractor opposite replacement air source

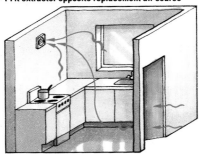

2 Place extractor near a cooker in a kitchen

Types of extractor fan

Many fans have an integral switch. If not, a switched connection unit can be wired into the circuit when you install the fan. Some types incorporate a built-in controller to regulate the speed of extraction and a timer that switches off the fan automatically after a certain interval. Axial fans can be installed in a window; and with the addition of a duct, some models will extract air through a solid or cavity wall (to overcome the pressure resistance in a long run of ducting, a centrifugal fan is required). To prevent backdraughts, choose a fan with external shutters that close when the fan is not in use.

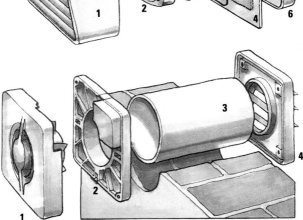

Window-mounted axial fan
1 Inner casing
2 Motor assembly
3 Interior clamping plate
4 Glass
5 Grille-clamping plate
6 Exterior grille

Wall-mounted axial fan
1 Motor assembly
2 Interior backplate
3 Duct
4 Exterior grille

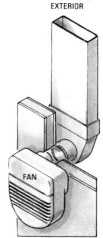

DUCTING TO EXTERIOR

FAN

Centrifugal fan

Choosing the size of a fan

The size of a fan, or to be accurate, its capacity, should be determined by the type of room in which it is installed and the volume of air it has to move.

A fan installed in a kitchen must be capable of changing the air completely ten to fifteen times per hour. A bathroom requires six to eight air changes per hour, and a WC six to ten. A living room normally requires about four to six changes per hour, but it's best to fit a fan with a slightly larger capacity in a smoky environment.

To determine the capacity of the fan you require, calculate the volume of the room (length x width x height) and multiply the volume by the recommended number of air changes per hour. Choose a fan that has at least that capacity (a slightly larger capacity is in fact best).

CALCULATING THE CAPACITY OF A FAN FOR A KITCHEN

Size			
Length	**Width**	**Height**	**Volume**
3.35m (11ft)	3.05m (10ft)	2.44m (8ft)	24.93cu m (880cu ft)
Air changes			
Per hour	**Volume**	**Fan capacity**	
15 x	24.93cu m (800cu ft)	= 374cu m per hour (13,200cu ft)	

73

FITTING AN EXTRACTOR FAN

Metal detector
Detect buried pipes
or cables by placing
a hired electronic
sensor against the
plaster.

**1 Hold panel in place
with plank**

2 Seal plate spigot

3 Insert duct in hole

4 Screw-fix grille

FITTING A WALL-MOUNTED UNIT

Satisfy yourself there is no plumbing or electrical wiring buried in the wall, using an electronic sensor (see left). Make sure there are no drainpipes or other obstructions.

Cutting the hole

Wall-mounted fans are supplied with a length of plastic ducting for inserting in a hole cut through the wall. Plot the centre of the hole and draw its diameter on the inside of the wall. Use a long-reach masonry drill to bore a central hole right through. To prevent the drill breaking through the brickwork or rendering on the outside, hold a stout plywood panel against the wall and wedge it with a strong plank supported by stakes driven into the ground (1).

Before cutting the brick, drill holes close together around the inner edge of the hole. With a cold chisel, cut away the plaster using the holes as a guide, then continue to cut away the brick-work (try to avoid debris falling inside a cavity wall). When you reach the centre of the wall, remove the panel, then use the same technique to finish the hole from the outside face.

Fitting the fan

Wall fans are mostly fitted in a similar manner, but check the instructions beforehand. Separate the components of the fan, then attach a self-adhesive foam sealing strip to the spigot on the backplate to receive the duct (2).

Insert the duct in the hole so that the backplate fits against the wall (3). Mark the length of the duct on the outside, remembering to allow for fitting the spigot on the outer grille. Cut the duct to length with a hacksaw. Reposition the backplate and duct in order to mark the fixing holes on the wall. Drill and plug the holes, then feed the electrical supply cable into the backplate before screwing it to the wall. Stick a foam sealing strip inside the spigot on the grille. Position it on the duct, then mark, drill and plug the wall-fixing holes. Use a screwdriver to stuff scraps of loft insulation between the duct and the cut edge of the hole (or use a sprayed expanding foam), then screw on the exterior grille (4). If the grille doesn't fit flush with the wall, seal the gap with mastic. Wire the fan according to the manufacturer's instructions, then attach the motor assembly to the backplate.

An extractor fan can only be installed in a fixed window. If you want to fit one in a sash window, then you will need to secure the top sash, in which the fan is installed, and to fit a sash stop on each side of the window in order to prevent the lower sash damaging the casing of the fan, should it be raised too far.

If you plan to install an extractor fan in a hermetically sealed double-glazing system, ask the manufacturer to supply a special unit with a precut hole, which is sealed around the edges, to receive the fan. Some manufacturers supply a kit for adapting a fan so you can install it in a window with secondary double glazing. It allows the inner window to be opened without dismantling the fan.

Cutting the glass

Every window-mounted fan requires a round hole to be cut in the glass. The size is specified by the manufacturer. It is possible to cut a hole in an existing window, but stresses in the glass will sometimes cause it to crack. And while the glass is removed for cutting, there is always a security risk, especially if you decide to take it to a glazier. All things considered, it is generally better to fit a new pane, which will be easier to cut and can be installed as soon as the old one has been removed.

Cutting a hole in glass is not easy, and you may find it's more economical to have it cut by a glazier. You will need

to supply exact dimensions, including the size and position of the hole. Use 4mm ($\frac{5}{32}$in) glass – either plain or obscured, to match the existing glazing.

Installing the fan

The exact assembly may vary, but the following sequence is a typical example of how a fan is installed in a window. Take out the existing window pane and clean up the frame, removing retaining sprigs and traces of old putty; then fit the new pane with the precut hole as you would any other window glass.

From outside, fit the exterior grille by locating its circular flange in the hole (1). Attach the plate on the inside, to clamp the grille to the glass. Tighten the fixing screws in rotation to achieve a good seal and even clamping force on the glass (2). Screw the motor assembly to the clamping plate (3). Wire up the fan in accordance with the maker's instructions. Fit the inner casing over the motor assembly (4). Switch on the fan to check that the mechanism runs smoothly, and that the backdraught shutter opens and closes automatically when the unit is switched on and off.

WARNING

Never make electrical connections until the power has been switched off at the consumer unit.

1 Place grille in the hole from outside

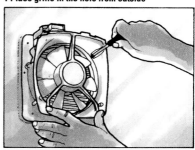

3 Screw the motor assembly to the plate

2 Clamp the inner and outer plate together

4 Attach the inner casing to cover the assembly

SEE ALSO
Details for:
Condensation   10-11

Window-mounted and wall-mounted fans are primarily intended for overall room extraction, but the most effective way to rid your kitchen of steam and greasy cooking smells is to mount an extracting hood – which is specifically designed for this purpose – directly over your cooker.

Where to mount the cooker hood

Unless the maker's recommendations indicate otherwise, an extracting hood should be positioned between 600mm (2ft) and 900mm (3ft) above a gas or electric hob or about 400mm (1ft 4in) to 600mm (2ft) above an eye-level grill.

Depending on the model, a cooker hood may be cantilevered from the wall or, alternatively, screwed between or beneath fitted kitchen cupboards. Some kitchen-unit manufacturers produce a special cooker-hood housing unit that matches the style of their cupboards (when the unit is opened, that operates the fan automatically). Most cooker hoods have two or three speed settings, and a built-in light fitting to illuminate the hob or cooker below.

Installing trunking

When a cooker hood is mounted on an external wall, air is extracted through the back of the unit into a straight duct passing through the masonry.

But if the cooker is situated against an interior wall, you'll need to connect the extracting hood to the outside by means of fire-resistant plastic trunking. The straight and curved components of the trunking – which simply plug into one another – form a continuous shaft running along the top of the cupboards fitted on the wall.

To fit the trunking, begin by plugging the female end of the first component over the outlet spigot attached to the top of the cooker hood. Cutting each of the components to length, as required, with a hacksaw or tenon saw, piece the rest of the trunking together, making the same female-to-male connections along the shaft. Some manufacturers print airflow arrows on the trunking to ensure each component is orientated correctly. If you should accidentally reverse a component somewhere along the shaft, then air turbulence may be created around the joint, reducing the effectiveness of the extractor. At the outside wall, cut a hole through the masonry for a straight piece of ducting and fit an external grille (see opposite).

Fitting a cooker hood

Recycling and extracting hoods are hung from wall brackets supplied with the machines. Screw-fixing points are provided for attaching them to the wall or to a cupboard. Cut a ducting hole through the wall, as for a wall-mounted fan (see opposite), and wire the cooker hood following the maker's instructions.

RECIRCULATION OR EXTRACTION?

The most important difference between one cooker hood and another is what they do with the stale air they capture. Some hoods filter out the odours and grease then return the air to the room. Others dump stale air outside through a duct in the wall, in much the same way as a wall-mounted extractor fan. Since the air is actually changed, extraction is the more efficient method.

In order to install an extracting hood, it is necessary to cut a hole through the wall then fit ducting and an external grille. Cooker hoods that recycle the air are much simpler to install – but they do not expel moisture from the room, nor do they filter out all the grease and cooking odours; it is also essential to clean and change the filters regularly to keep a recirculation hood working at peak efficiency.

Recirculation hood returns filtered air to room

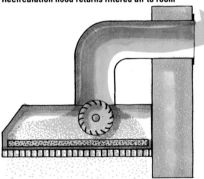

Extraction hoods suck air outside via trunking

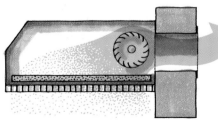

Alternatively, air is extracted through ducting

Running trunking outside
When a cooker is placed against an inside wall, run plastic trunking from the extractor hood along the top of wall-hung cupboards.

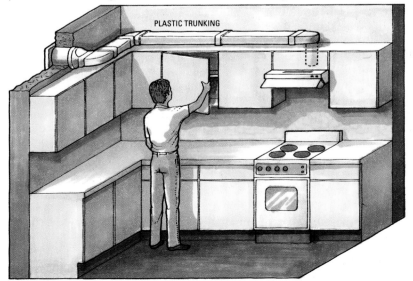

PLASTIC TRUNKING

HEAT-RECOVERY VENTILATION

It is estimated that more than half the energy produced by burning fossil fuels is used simply to keep our homes warm. Although the installation of efficient insulation reduces heat loss to a minimum, a great deal of heat is still wasted as a result of necessary ventilation.

Heat-recovery ventilators are designed to balance the requirements of conserving energy and the need for a constant supply of clean fresh air. They range in size from compact airbrick-size units for continuous low-volume ventilation of individual rooms to whole-house ducted systems.

How heat-recovery ventilators work

This type of ventilator contains two low-noise electric fans. Stale air from the interior is extracted by one fan through a highly efficient heat exchanger. This absorbs up to 70 per cent of the heat that would otherwise be wasted, and transfers it to a flow of fresh air drawn into the room by the second fan. Since the two airflows don't mingle, odours and water vapour are not transferred along with the heat.

Self-contained heat-recovery ventilators can be fitted in exteriors wall or windows. The extraction unit of a larger ducted systems is usually mounted in the loft or a cupboard, and the cooker hood is normally connected to a whole-house system.

Designed for straightforward regular maintenance, both the air filters and the heat exchanger can be removed easily and washed in soapy water.

DEHUMIDIFIERS CONTROL CONDENSATION

To combat condensation, you can either remove the moisture-laden air by ventilation or warm it so that it is able to carry more water vapour before it becomes saturated. A third possibility is to extract the water itself from the air, using a dehumidifier.

A dehumidifier works by drawing air from the room into the unit and passing it over a set of cold coils, so that the water vapour condenses on them and drips into a reservoir. The cold but now dry air is then drawn by a fan over heated coils before being returned to the room as additional convected heat.

The process is based on the simple refrigeration principle that gas under pressure heats up – and when the pressure drops, the temperature of the gas drops too. In a dehumidifier, a compressor delivers pressurized gas to the 'hot' coils, in turn leading to the larger 'cold' coils, which allow the gas to expand. The cooled gas then returns to the compressor for recycling.

A dehumidifier for domestic use is built into a cabinet resembling a large hi-fi speaker. It contains a humidistat that automatically switches on the unit when the moisture content of the air reaches a predetermined level.

When the reservoir is full, the unit shuts down in order to prevent overflowing, and an indicator lights up to remind you to empty the water in the container.

When a dehumidifier is installed in a damp room, it should extract the excess moisture from the furnishings and fabric within a week or two. After that, it will monitor the moisture content of the air to maintain a stabilized atmosphere. A portable version can be wheeled from room to room, where it is plugged into a standard wall socket.

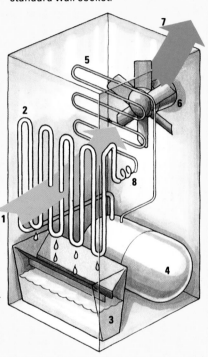

Working components of a dehumidifier
The diagram illustrates the layout of a typical domestic dehumidifier.
1 Incoming damp air
2 Cold coils
3 Water reservoir
4 Compressor
5 Hot coils
6 Fan
7 Dry warm air
8 Capillary tube where gas expands

Heat-recovery ventilation unit
The diagram shows the layout of a typical wall-mounted heat-recovery ventilator.
1 Stale air from room
2 Stale-air exhaust
3 Fresh-air supply
4 Warmed fresh air
5 Heat exchanger
6 Induction fan
7 Extractor fan

Fitting a heat-recovery ventilator

Site a heat-recovery ventilator on an external wall, close to the ceiling and in a position where it will extract air most efficiently. Although typically set into a standard solid or cavity wall, you can install one in a thicker wall with the aid of a telescopic metal sleeve. The more compact units can be fitted in windows and single-leaf walls.

After marking out the aperture, cut away the brickwork, as described in the section on fitting a standard extractor fan. Fit the unit into the hole and make good the masonry and plasterwork, sealing any gaps around the unit with a gun-applied sealant. Finally, wire up the controls of the ventilator, following the manufacturer's instructions.

Aggregate
Particles of sand or stone mixed with cement and water to make concrete.

Architrave
The moulding around a door or window.

Alkali-resistant primer
A primer used to prevent the alkali content of some building materials attacking subsequent coats of paint.

Arris
The sharp edge at the meeting of two surfaces.

Balanced flue
A ducting system which allows a heating appliance such as a boiler to draw fresh air from and discharge gases to the outside of a building.

Batt
A short cut length of glass- or mineral-fibre insulant.

Batten
A narrow strip of wood.

Blind
To cover with sand.

Blown
To have broken away, as when a layer of render has parted from a wall.

Buttercoat
The top layer of render.

Casing
Timber lining of a doorway.

Cavity tray
A sloping surface built in the void of a cavity wall to shed water to the outer leaf.

Cavity wall
A wall of two separate masonry skins with an airspace between them.

Chamfer
A narrow flat surface on the edge of a piece of wood – it is normally at an angle of 45 degrees to adjacent surfaces. or To plane the angled surface.

Chase
A groove cut in masonry to accept pipes or cable. or To cut such grooves.

Consumer unit
A box which contains the fuses or MCBs that protect the household circuits. It also incorporates the main switch which cuts the power to the whole building.

Core drill
A hollow drill bit tipped with saw teeth for cutting large-diameter holes in masonry.

Damp-proof course – DPC
A layer of impervious material which prevents moisture rising from the ground into the walls of a building.

Damp-proof membrane – DPM
A layer of impervious material which prevents moisture rising through a concrete floor.

Drip groove
A groove cut or moulded in the underside of a sill to prevent rainwater running back to the wall.

Efflorescence
A white powdery deposit caused by soluble salts migrating to the surface of a wall or ceiling.

Expansion pipe
An expansion pipe drains into the storage cistern in the roof to allow for expansion of heated water in the hot-water cylinder and to vent air from the plumbing system. Also known as vent pipe.

External wall insulation
Thermal insulating material which is fixed to the outer surface of a house in order to prevent transmission of heat to the outside.

Fascia board
A strip of wood which covers the ends of rafters and to which external rainwater guttering is fixed.

Flaunching
A mortared slope around a chimney pot or at the top of a fireback.

Footing
A narrow concrete foundation for a wall.

Frass
Powdered wood produced by the activity of woodworm.

Furring strips
Parallel strips of wood fixed to a wall or ceiling to provide a framework for attaching panels.

Fused connection unit
A device for permanently connecting electrical cable to flex – or sometimes another cable – from an appliance. The unit is protected by a cartridge fuse.

Hardcore
Broken bricks or stones used to form a sub-base below foundations, paving, etc.

Jamb
The vertical side member of a door or window frame.

Key
To abrade or incise a surface to provide a better grip when gluing something to it.

Ladder stay
A device which holds the top of a ladder away from a wall.

Lath and plaster
A method of finishing a timber-framed wall or ceiling. Narrow strips of wood are nailed to the studs or joists to provide a supporting framework for plaster.

Long-reach masonry drill
A drill with an extended shank for boring through thick walls.

Marking gauge
A woodworking tool designed to scribe a line parallel to a straight edge.

Mastic
A non-setting compound used to seal joints.

Moisture-vapour permeable
Used to describe a finish which allows timber to dry out while protecting it from rainwater.

Mono-pitch roof
A roof which slopes in one direction only.

Mortar board
A 1m (3ft) square of exterior-grade plywood upon which mortar or plaster is mixed with water.

Mullion
A vertical dividing member of a window frame.

Oxidize
To form a layer of metal oxide as in rusting.

Paint stripper
A chemical which softens old paint so that it can be removed from a surface.

Paint system
Layers of paint applied to a surface, consisting of primer, undercoat and top coat.

Pallet
A wooden plug built into masonry to provide a fixing point for a door casing.

Panelling
Strips of solid timber or man-made boards used to line the face of a wall as a decorative finish.

Plugging chisel
An all-metal chisel with a flat narrow bit (tip) for cutting out old pointing between bricks or blocks.

PTFE
Polytetrafluorethylene – used to make tape for sealing threaded plumbing fittings.

Render
A thin layer of cement-based mortar applied to exterior walls to provide a protective finish. or To apply the mortar.

Sash stop
A metal stud screwed to the framework of a sash window to prevent it being opened further than required for ventilation.

Scaffold tower
A tall structure built from metal poles and boards.

Scratchcoat
The bottom layer of render.

Screed
A thin layer of mortar applied to give a smooth surface to concrete etc.

Screed batten
A strip of wood fixed to a wall as a guide to the thickness of plaster or render.

Stile
A vertical side member of a door or window sash.

Work platform
A structure built from scaffolding or trestles and boards, used to gain access to walls and ceilings for decoration or repair.

Page numbers in *italics* refer to photographs and illustrations

flaunching 77
float glass 21
floor paint 40, 41
flush joints 33; *33*
footing 77
frass 77
Frenchman 33; *33*
fungicidal preserver 15, 16, 32
fungicidal solution 32
furring strips 50, 77; *50*
fused connection unit 77

G

gable end 45
Georgian wired glass 21
glass
  buying 22
  cutting 22-3, 74; *22, 23*
  drilling 23; *23*
  fitting 24; *24*
  measuring 22
  removing 17, 24; *24*
  types 21
glass-repair tape 55
glazed roofs 55
guttering 8, 56-8; *56, 57, 58*
  unblocking 57

H

hardcore 77
hardener, wood 15; *15*
heat-recovery units 76; *76*
hinges
  fitting 26-7; *26, 27*
  swapping 28; *28*
hipped end 45
hips, roof 44; *44*
hopper head 56; *56*
horns, door 26
hot-water cylinder 64

I

infestation 44
injection process, DPC 13; *13*
insulated plasterboard 63
insulation
  attic room 61; *61*
  cavity wall 63; *63*
  ceiling 62; *62*
  flat roof 62; *62*
  hot-water cylinder 64; *64*
  loft 59-60; *60*
  pipes 60, 64; *64*
  sloping roof 61; *61*
  tank 60; *60*
  types 59
  walls 62-3; *63*

J

jamb 77
joist 43, 77

K

key (grip) 77

L

laced valley 47
ladder stay 77
lagging 60, 64; *60, 64*
laminated glass 21
lath and plaster 77
lead, patching 55
leaks, patching 14
lime 36
loft insulation 59, 60
long-reach masonry drill 77
loose-fill insulation 59, 60; *60*

M

marking gauge 77
masonry
  cleaning 32-3
  cracked 34
  paint 40, 41
  painting 40-41; *41*
  waterproofing 34
mastic 10, 77; *10*
mastic asphalt 50
metal detector 74
moisture meter 12, 29; *12*
moisture-vapour permeable 77
mono-pitch roof 77
mortar 36, 37
  additives 37
  mixes 37
  mixing 38; *38*
mortar board 77
mortar dyes 33
mortar joints 33; *33*
mould growth 15, 32
mycelium 15

N

nailable plug 63
non-reflective glass 21

O

organic growth 32
oxidize 77

P

PTFE 77
PVC glazing 67
paint
  preparing 40
  types 41
  stripping 33; *33*
paint stripper 77
paint system 77
painting
  masonry 40-41; *41*
  techniques 40-41; *41*
pallet 77
panelling 77
pantile vent 71; *71*
parapet 51; *51*
parapet gutter 56
patterned glass 21
pebbledash 35; *35, 39*
penetrating damp 7-9
physical DPCs 12
pipe insulation 60, 64; *64*
pitched roof 43; *43*
planning permission 19
plasterboard 63; *63*
plastic corrugated sheeting 49
plastic-film glazing 66, 67; *66*
plastic-sheet glazing 67
plugging chisel 77
pointing 8, 33
polycarbonate glazing 67
polyester film 67
polystyrene 67
polyurethane 14; *14*
porous tubes 12; *12*
Portland cement 36
preservers, timber 16; *16*
preservative tablets 16; *16*
purlin roof 43
putty fixing 24; *24*

R

radiator-heat reflector 64; *64*
rafter 43, 78
raked joints 33; *33*
render 36, 77
  applying 39; *39*
  defective 8
  one-coat 39
  repairing 35, 39; *35*
  textured 39
  two-coat 39
repair tapes, self-adhesive 53, 55; *53, 55*
repointing 33; *33*
ridge board 43
ridge tiles 49; *49*
rising damp 7, 9

79